PRINCIPLES OF STATISTICAL TECHNIQUES

(SECOND EDITION)

PRINCIPLES OF STATISTICAL TECHNIQUES

A First Course, from the Beginnings, for Schools and Universities

WITH MANY EXAMPLES AND SOLUTIONS

BY

P. G. MOORE

Professor of Statistics and Operational Research
London Graduate School of Business Studies

(SECOND EDITION)

CAMBRIDGE UNIVERSITY PRESS

CAMBRIDGE

LONDON NEW YORK NEW ROCHELLE

MELBOURNE SYDNEY

CAMBRIDGE UNIVERSITY PRESS
Cambridge, New York, Melbourne, Madrid, Cape Town, Singapore, São Paulo, Delhi

Cambridge University Press
The Edinburgh Building, Cambridge CB2 8RU, UK

Published in the United States of America by Cambridge University Press, New York

www.cambridge.org
Information on this title: www.cambridge.org/9780521076319

First published 1958
Reprinted (with solutions) 1964
Second edition 1969
Reprinted 1974
First paperback edition 1979
Reprinted 1980
Re-issued in this digitally printed version 2008

A catalogue record for this publication is available from the British Library

Library of Congress Catalogue Card Number: 70–85731

ISBN 978-0-521-07631-9 hardback
ISBN 978-0-521-29055-5 paperback

CONTENTS

PREFACE

During recent years the importance of the subject of statistics has become increasingly recognised and it is now studied not only by statistical specialists but by students from many different disciplines. It has also been recognised that the subject is a suitable one for all levels of educational activity in that it can provide, at quite an early stage, a unifying link between the theoretical and practical sides of many forms of scientific training. This book is an attempt to put across the main principles of statistical methods to those students who are fundamentally interested in the practical applications of the subject, but are not so much concerned with the philosophical bases of the concepts used.

The choice of what to include and what to omit has been difficult, and this second edition of the book includes some new material whilst omitting some of the earlier material. Primarily the aim has been to give a selection of the more commonly used tools and not to provide a complete set of statistical tools for use in each and every situation. The student is then left in the position where he should be able to appreciate what further tools are needed and he can usefully profit from a reading of the more advanced books on the subject that are available. To have included every technique in common use would have lengthened the present book very considerably and destroyed a greater part of its planned utility. Hence experienced readers must not be surprised if some of their trusted favourites are missing from the pages that follow.

To some extent the choice of topics has also been influenced by the desire to keep the standard, and the amount of mathematics down to a minimum. The basic mathematics required, with the exception of one or two symbols that are explained in the text, and small portions of some of the later chapters, is roughly that of Ordinary Level in the General Certificate of Education. Even this standard is not necessary for studying the earlier chapters and it is quite feasible for the book to be taken by schools in portions over a number of years. At a university it would most likely be a suitable basis for a one-year course of lectures to scientists or social studies students who are not mathematical specialists.

There are numerous examples in the text, most of them requiring a certain amount of calculation. One problem has been to decide the degree of accuracy to which these calculations should be carried. Readers will probably have a wide variety of computational aids at their command, ranging from slide-rules to electronic computers, and having very different accuracies. All results quoted in the text are accurate to the number of figures given, but this accuracy will not always be attained with four-figure logarithms and even less often with a slide-rule. The final chapter is the only one where serious difficulties are likely to occur. In this chapter the common form of slide-rule is definitely not accurate enough, although reasonable results should be obtained with four-figure logarithms.

The data given in the examples and exercises have been drawn from a wide range of reports, magazines, journals and books. In many cases the original data have been greatly tampered with, and reductions, groupings or simplifications have been made before using the data to illustrate a particular point. In these cases the source has not been given for fear of misrepresentation of the original author's intentions. Where the data are substantially in the original form, due acknowledgment has been made. Tables 9.4, 12.1 and 12.2, giving the normal, χ^2, and t distributions, have been extracted from rather fuller tables in *Biometrika Tables for Statisticians* by kind permission of Professor E. S. Pearson, the editor of *Biometrika*.

It is a pleasure to acknowledge the great help received from many quarters in the preparation of this book. In particular, Dr C. L. Mallows kindly read and commented on the first edition, whilst Mr S. D. Hodges commented on the second edition. My wife has given me a great deal of help with both editions, particularly with the proof-reading. A number of suggestions from readers of the first edition have been incorporated in this new edition, and comments from readers of this second edition would likewise be welcomed.

P.G.M.

December 1968

1

THE SCOPE OF STATISTICS

1.1 Broadly speaking, statistics is the numerical study of a problem. Unless the problem can be reduced to quantities measurable on a scale or capable of being expressed as a number, it is impossible to make a statistical study of it. Statistics is not, however, just concerned with the counting of individuals or the measuring of items. Its ramifications are far wider than that, and include the study of what are the right figures to collect and the correct interpretation to be placed on them. The politician trying to envisage the effects of different forms of taxation needs to know the estimated yields of each form of taxation proposed; and the local town councillor must be able to appreciate how the local rate is split up into various headings. The citizen today is deluged with White papers, Economic surveys and a multitude of reports not only from the Government but from banks, insurance companies and industrial firms, all of which present, and argue from, a mass of statistical data. An understanding of statistics and the treatment of numerical data is therefore essential and only by a patient study of the part played by figures in such reports can good decisions be made and policies understood and, if need be, criticised. It is necessary to recognise, moreover, the power and the limitations of statistical arguments, to learn how to obtain the full information from a set of figures and how to avoid the pitfalls which await the unwary. If the figures are worth analysing at all they are surely worth the form of analysis that yields the maximum amount of information. There are a large number of statistical tools, and to use a steam-hammer where a light tap is required would be not only wasteful but often misleading. No one universal rule can be made and both knowledge and experience must be gained if the best possible results are desired.

The whole subject of statistics has taken tremendous strides since the beginning of the century, and the last war gave a big impetus to the further study of statistical methods, since the use of such methods often led to large savings of time, materials and personnel. Before a study of the basic methods of the subject, a number of examples will be given to illustrate some of the many fields of its application. Whilst these examples are in no way exhaustive they are nevertheless instructive in that they give some idea of the multitude of problems that confront the statistician.

1.2 Government. For effective government and the shaping of policy it is necessary to have accurate statistical knowledge of the exact composition of the population. From this need has sprung in the first place the compulsory registration of births, marriages and deaths, and secondly the census which was inaugurated at the time of the Napoleonic wars and is normally taken in Great Britain once in every ten years. The information derived from these sources gives an instantaneous picture, as it were, of the population divided up by age, sex, housing and so on. This information is needed in order to see, for example, how many schools are required, how many workers there are, and how many persons are pensioners and hence no longer productive workers; estimates can then be made of important quantities, such as the working population ten years hence. By collecting details of housing conditions (number of rooms, washing facilities and so on) the census also gives valuable information about the social well-being of the nation.

Table 1.1. *Acreage in Great Britain utilised for agriculture in June 1967*

Type of agriculture	Area in thousand acres
Arable land under crops	12,354
Arable land under grass	5,934
Permanent grassland	12,328
Rough grazing	17,639

The well-being of a nation, however, depends not only on the population but also on the land and such natural resources as minerals, forests and livestock, and on capital equipment in the form of buildings, factories and machinery. To obtain information on all these subjects the various government departments make large-scale surveys at regular intervals. Much of the material obtained by the government is published in the *Monthly Digest of Statistics*, from which the information in table 1.1 is taken. This publication is probably in your local library. A study of it will give you some idea of the vast amount of information used by the government in making decisions concerning the day-to-day policy of the country.

1.3 Industry. The wealth of the country does not depend primarily on the government. It has to be generated elsewhere, and industry is one of the main sources of that wealth. The production of marketable goods of adequate quality provides innumerable statistical problems of which the following are typical examples.

A manufacturer is making electric light bulbs and according to the design and specification the bulbs should burn for 2,000 hr. Due to slight differences in manufacture the bulbs will not all have exactly the same length of life, but will vary amongst themselves. A batch of 500 bulbs has just been produced and the manufacturer is going to make tests in order to see whether the bulbs are up to the standard and have a burning life of at least 2000 hr. Quite clearly he cannot test every bulb by measuring the time it takes to burn out for there would then be none left for sale. Hence it is essential to use some method whereby a few of the bulbs are examined and inferences made from them about the whole batch. If the whole batch could be examined it would be possible to make a categorical statement such as 'the bulbs all have a life of at least 2,000 hours'. As only a few of the bulbs can be examined the statement must take the form 'the bulbs almost certainly all have a life of at least 2,000 hr.' or 'the bulbs are very unlikely to have a life of at least 2,000 hr.', or something in between these two statements. But provided that the selection and examination of the bulbs is in accordance with the principles given later, then the latter forms of statement may give as much information as is required at a fraction of the cost of a complete examination, which in any event is an impossibility in this case.

A similar situation would arise in the determination of the breaking strength of a batch of steel wires, where a test destroys the wire, and the desired information must be obtained by performing tests on a selection of the wires. Such a procedure is necessary in order to check whether the quality of the product is being maintained, as a failure could be dangerous and might well result in a falling-off of sales.

Industrial problems also arise in the trial of new processes, as in the following example. Two batches of cloth are made by different processes, A and B, and the resistance of the cloth to acid is tested by taking four pieces as samples from each batch and measuring the length of time for which they resist the acid. The results, in hours, of a typical experiment were as follows:

	Process A	Process B
Piece 1	40·9	43·8
2	40·7	41·7
3	41·6	43·1
4	41·2	44·2

Process B is a new process and the question is whether it is better than Process A? It is true that all four results for B are higher than those of A, but only just. Notice that the four values for B vary much more amongst themselves than do those for A. This suggests that the samples from B are not so consistent as those from A. The final decision must reconcile all these factors and this type of problem is analysed in detail later in this book.

1.4 Road Safety. Industrial goods have to be moved to the places where they are needed and much of the transport is by road. In recent years there have been more and more accidents on the roads, with the result that new safety measures are constantly being devised. The only sound basis for judging the relative effectiveness of suggested precautions for the reduction of road accidents is a statistical one. Thus a common form of statement is that road junction X is much more dangerous than road junction Y. A statement of this sort cannot be based merely on the number of accidents that occur at the two junctions. One junction may carry much more traffic or may have more traffic at dusk, which is a peak period for accidents. Or again, at one junction all the traffic may go straight across whereas at the other a large proportion may make a right turn.

Other common assertions are that high speeds are the most frequent cause of accidents, or that different parts of the country have different accident rates, or that one form of road surface is more dangerous than another. To substantiate these statements a considerable amount of statistical evidence is needed, since the opinions of individuals are widely conflicting, and only by a careful numerical study of the various factors involved is it possible to arrive at an impartial decision.

Statistical work is also necessary in solving the problem of traffic congestion in towns. To make vast and expensive alterations to the system of traffic control without making sure the alteration will have the desired effect is both useless and wasteful. For instance, the common type of roundabout will take only a certain amount of traffic per hour, and any attempt to force more traffic to use it will only result in large-scale congestion. Hence before installing a roundabout a detailed study of the volume of traffic coming into the junction at various times of the day must be made. If this is not done a hoped-for improvement could have the opposite effect.

London and most other cities have large and complicated networks of public transport and, although such networks may appear somewhat haphazard to the uninformed, there is scope for an enormous volume of statistical work in the background. The organisers want to know how people get to work, or to the shops, or to places of entertainment, and how long it takes them; they want to know the effect of alterations in fares on total receipts, and on the pattern of the journeys. All these things and many more must be studied in order to choose the most useful routes, and arrange the most convenient time-tables, consistent with keeping the running costs as low as possible.

1.5 Insurance. The citizen comes into even more direct contact with statistics in the field of insurance. For example, Mr A insures his house against damage by fire by paying an annual sum or premium to an insurance company. The amount of this premium is not arbitrary, but is governed by the numbers of fires that do occur in houses of similar type and the amount of damage they cause. There are factors that enable the financial risk of a fire at Mr A's house to be directly assessed. Again to assess the correct premium to charge for insuring a motor car requires knowledge of the claims made on that type of car, for the particular age of the driver, his occupation, previous claim record and so on. All this information needs to be collated and carefully assessed so as to yield the important factors that determine the correct premium to charge.

An alternative form of insurance policy is that for an annuity.

According to the prospectus of one large assurance office they will grant an annuity of £11.30 to a man now aged 60 if he will pay them £100. This means that if Mr *B*, now aged 60, were to pay over to the office the sum of £100 they would pay Mr *B* the £11.30 every year for the rest of his life. If Mr *B* dies in two years' time he will receive two payments, but if he lives to be a centenarian he will receive forty payments. In order to determine the amount of the annuity it can offer the assurance office must have accurate information on the number of years a man aged 60 is likely to live, and on the rate of interest they may expect to earn on invested money. A study must therefore be made of the distribution of ages at death of all men over 60 in the past, and examination made of the rates of interest that are obtainable on various forms of investment. Then the rate of annuity is calculated by spreading the risks over all men aged 60 who buy annuities.

1.6 Market Research. All the goods and services available to the public have to be shown or demonstrated to them in order that people shall know of their existence. The demand has to be measured and the reactions of the public noted for future developments. Imagine that a firm is putting on the market a new type of washing machine. Obviously it is desirable to know what will be the approximate demand for the machine. If the demand is only going to be of the order of 100 machines a week it is very unprofitable to set up plant and machinery designed to turn out washing machines at the rate of 5000 a week. The concept of gauging the demand for a product is the basic problem in market research. In its simplest form it is answered by questioning a proportion of households on their intentions, that is, the likelihood of their purchasing a washing machine of this kind. A similar procedure is followed by the B.B.C. in its Listener research organisation, which continuously investigates the popularity or otherwise of its programmes. The statement that 8,000,000 people listen to such and such a programme does not mean that every person in Great Britain has been questioned. In practice only a few thousand will have been questioned, but care will have been taken to see that these are representative of the whole population, so that valid deductions about the listening habits of the whole population may be made.

1.7 The fields of application of statistics are boundless and cannot all be mentioned. In agricultural work the decision as to whether one variety of wheat is better than another will be made on the results of a carefully planned series of statistical experiments. In biology the effects of two drugs on rabbits, may be compared and the effects of the two drugs assessed on some numerical scale. Again, if a company makes steel bars of a specified length, the bars turned out in practice will inevitably vary in length. Management must determine the average length for which the process should be set so that the total scrap—resulting from trimming long bars plus that resulting from scrapping the short ones—is at a minimum.

All these examples show how statistical methods are essentially a guide to action or decision of some form. It replaces a vague personal impression such as 'I think that Drug A is better than Drug B' by a well-defined and clear-cut statement of the form 'Drug A is 40 per cent more powerful than Drug B'. Everyone argues from general impressions, and the art of conversation would be very difficult if sweeping statements such as 'aeroplanes are more dangerous than cars' were inadmissible without supporting statistical evidence. Nevertheless general impressions are often misleading and sometimes untrustworthy. For example, people are apt to say that they are always getting wrong numbers when using the telephone. A statistical count would, in all probability, show that the proportion of wrong numbers was really very small. As is so often the case, the times when things go wrong are remembered but the numerous times when all goes well are forgotten.

The examples of the applications of statistics given above illustrate the usual pattern of a statistical investigation in which there are four phases, namely:

(*a*) Statement of the problem to be investigated.

(*b*) Collection of the data needed, either from available sources or by performing experiments.

(*c*) Analysis of the experimental results or data.

(*d*) Interpretation of the results of the analysis.

Thus to see how many schools will be needed in Manchester in 1980 is to state the problem as in (*a*). Next, from census figures and local figures relating to movements of families in and out of the city, together with a forecast of birth rates, the numbers of children in various age-groups expected to be in Manchester in 1980 must be obtained. This constitutes (*b*). The data must now be sifted down in order to produce the comprehensive figure required for stage (*c*) and finally in stage (*d*) the decision is made on the number of schools required.

Or again the insurance company in section 1.5 wishes to find how much annuity it can grant for £100 to a man aged 60. This is the problem under (*a*). Next, the company collects together all the available information as to how much longer men aged 60 have lived in the past, together with the rates of interest at which the company is likely to be able to invest its money. This is stage (*b*) and leads to stage (*c*), the sifting of this information to give the required figures. Since there may be alternative estimates of the annuity from the analysis in (*c*) the interpretation of the results under (*d*) will require the exercise of judgment and experience in order to decide what annuity can be offered.

In the remainder of this book these subdivisions of any investigation will be discussed at some length, and the powerful aid that statistical methods can give to clear thinking and rational decisions will be demonstrated. Since each problem that turns up will be slightly different from the next one, practice is essential in order to acquire facility, and the student is urged to work through as many of the examples and exercises as is possible.

2

THE COLLECTION OF DATA

2.1 In the opening stages of a statistical inquiry the investigator will need to collect a large amount of raw material or data from which to extract the quantities relevant to the purposes that he has in mind. Thus the market research investigator may have to collect data on consumer preferences after four different sets of advertisements have appeared; the botanist may have to spend several days in a grassland area counting the number of shoots of *Solidago glaberima* per square foot; the traffic investigator may have to count traffic at a busy crossing; the agriculturalist may have to collect data concerning the quantity of fertiliser applied to wheat crops on all the farms in Sussex. The method and care given to the collection of this raw material is important. The strength of a chain lies in the strength of its weakest link and it is useless to reach intricate conclusions from insufficient or inaccurate data. Before making any form of elaborate analysis it is essential to know the limitations and accuracy of statistical material and to be aware of the kind of errors that can arise. In this chapter two of the most common sources of data will be considered in some detail. These are:

(i) *Questionnaires.* The data here are obtained by forms designed by the statistician and completed by the general public.

(ii) *Observations.* The data here are collected by the investigator himself recording the results of a series of observations but not necessarily relying on the public at large for his information.

2.2 Questionnaires and their completion have to some extent become a part of the daily life of the citizen in this country, though there have been for some time special forms concerning each citizen's history. For example, his birth must be registered at a local registrar's office, within 42 days, on a form somewhat similar to that shown in fig. 2.1. This registration is required by British law, and on marriage and at death similar types of form have to be completed. These forms provide the raw material for a large number of studies made by the government into the size, age-distribution and marital status of the population. The record of a birth, marriage or death can be inspected at Somerset House in London, where the records have been kept since this type of registration became compulsory in 1874. It is sometimes necessary to prove one's age (in order to sit for certain examinations, for example, or to receive a legacy conditional on being twenty-one) and the birth certificate provides a ready means for this purpose. However, these records of births, marriages and deaths are not always by themselves sufficient for many statistical problems. They give no idea, for example, of the population of a town at a given time or of the occupations followed by its residents. To answer such questions some further source of information is necessary and some more continuous check on movements would therefore be necessary. This was attempted during the last war by means of identity cards which were useful also in operating a fair system of rationing. In normal times, however, this continuous check is regularly made by carrying out a census.

Birth in the district of......................in the County of..............

When born	Name	Sex	Name and surname of father	Name and maiden surname of mother	Profession of father	Date of registration	Signature of registrar

Fig. 2.1. Abbreviated form of a birth certificate

2.3 A census is a very comprehensive affair and in peace time is carried out simultaneously throughout the length and breadth of Great Britain every ten years. The last complete census was held on 23 April 1961, when the head of every household was responsible for the correct filling up of a schedule asking for certain particulars for every member of his household. There were in all thirty-six questions on the schedule but many of them did not apply to everyone. The questions fell, broadly speaking into the following categories:

1. Name.
2. Sex.
3. Age.
4. Whether married, etc.

5. Usual residence. 6. Place of birth and nationality.
7. Amount of education and qualifications held.
8. Employment at time of census.
9. Profession or trade or if at school.
10. Particulars of house, e.g. number of living rooms.

Great care is taken to ensure that census schedules are filled in correctly and that everybody completes a schedule. Many people are away from their normal place of residence on the day of the census. They may be at work elsewhere, or travelling, or at sea, or on holiday, but it is essential that they are all brought into the census. In 1961 some 50,000 specially trained enumerators were employed to deliver the blank schedules, to explain to householders exactly what had to be done and later to collect the completed schedules from each household. Strenuous efforts were made to capture the interest of the people by means of broadcasts and newspaper articles, and to drive home the importance of correct compilation of the schedules. As a result of all this work the final results can be taken as being absolutely reliable for most practical purposes.

2.4 As a census is so expensive it can only be carried out at long intervals. The information, however, can rapidly become out of date and hence policy decisions may be endangered by innacurate data. In 1966 a sample census was held on 24 April (i.e. halfway between the usual census dates) to obtain information on similar lines to the 1961 census, but from only 10 per cent of the population. Although only a fraction of the total population was asked to supply information, provided the selection of that fraction is carefully controlled, valuable information can be obtained, as will be demonstrated later in this book. Furthermore, there is a considerable saving in processing time, thus enabling the results to be available that much closer to the date of origination of the information. The summary tables for the 1961 census were published in 1966; those for the 1966 sample census were published in 1967.

2.5 The information in a census is obtained from the answers to the questions on the schedules. Clearly the efficiency of the census depends largely on how those questions are framed, for bad questions can produce wrong information. This applies to all inquiries made by questionnaire, and in the design of any statistical form there are certain rules which must be followed if reasonable results are to be obtained. Briefly these rules are:

(*a*) The form should be as concise as possible and there should be the minimum number of questions necessary to obtain the required information.

(*b*) The questions should be simple, and unambiguous in their possible interpretation.

(*c*) Questions that are likely to arouse strong feelings and hence attract inaccurate answers should be avoided. For example, asking a man if he has any physical deformity is quite likely to produce an incorrect answer.

(*d*) The form should be made as attractive as possible to the eye by means of a suitable layout and clear type.

(*e*) When asking for confidential information in a voluntary inquiry the person's name should not be put on the form unless it is essential. This precaution is likely to produce accurate replies to the confidential questions.

To illustrate the method of framing questions suppose that it is desired to make an estimate of the total mileage driven in this country in a year. A number of different methods have been suggested and the first method to be tried was based on the total consumption of petrol in the country. A suggested alternative method was to have a questionnaire completed by every motorist when renewing his or her driving licence. The form (fig. 2.2)

Name: Address:
Age:
Employment:

	Bus or lorry	Car	Motor cycle
Approximate mileage you have driven in past week			
Approximate mileage you have driven in past month			
Approximate mileage you have driven in past year			

Fig. 2.2. Specimen form

appears simple enough at first, but in order to assess its effectiveness some of the above-mentioned criteria will be applied. The form is fairly concise and the number of questions asked, thirteen, is reasonably low, especially as it is unlikely that all thirteen questions would have to be answered by any one person. Thus condition (*a*) seems reasonably well satisfied. The form cannot, however, be said to be free from ambiguity. First it asks for the name and does not make it clear whether the christian name of the driver is required, as it is if the sex of the driver is to be deduced. Secondly it is probably unnecessary to ask for the full address, as in many cases only the town and county of residence are required for the purpose of further analysis. Many drivers will omit the county unless specifically asked to put it in, and thereby give a great deal of extra work for the investigators. Next the question asking for the age is not precise as the age given by the driver may be (i) age last birthday, (ii) age next birthday, or (iii) age to nearest birthday. A person who gives his age as 19 years would be, under method (i), between 19 and 20, under (ii), between 18 and 19, whilst under (iii), between $18\frac{1}{2}$ and $19\frac{1}{2}$. The question would be better if it asked either the date of birth or the age in one of the three categories just mentioned. Further trouble can be anticipated from the answers to the questions on employment. Differences of personal opinion may lead to the same job being called by a multitude of names. Employment is probably the wrong word to use in this question. For instance, a coal mine employs large numbers of men who do no actual mining and a clerk working in the pay office of a mine might describe his employment as coal mining. His real work is, however, that of a pay clerk and it is probably better to ask for a person's occupation rather than his employment. Lastly the question on mileage fails to make it clear that what is required is the actual mileage driven, whether the vehicle was owned by the person in question or not, and not the mileage traversed as a passenger. Thus considerable improvements could be effected under conditions (*b*).

This questionnaire is unlikely to arouse strong feelings, nor does it ask for very confidential information, and so can be said to fulfil conditions (*c*) and (*e*). Its success under (*d*) is to some extent a matter of judgment in layout, bearing in mind the cost of printing or reproduction. Clear bold type with a minimum of small type or footnotes should be aimed at. Taking all these criticisms into account the form can now be re-designed as in fig. 2.3. To obtain the best results some sort of accompanying letter, explaining why the survey is being carried out and how the driver can help by completing the form, will be necessary.

Name:	Mr/Mrs/Miss
Age last birthday:	Town of residence:
What is your occupation?	County of residence:

	Bus or lorry	Car	Motor cycle
Approximate mileage you have driven in past week			
Approximate mileage you have driven in past month			
Approximate mileage you have driven in past year			
(Note that the mileage to be entered is the mileage that you have actually driven whether you owned the vehicle or not. Travel as a passenger is to be omitted.)			

Fig. 2.3. Re-designed form

2.6 The foregoing discussion has shown that the designing of any questionnaire entails a great deal of thought and that it is essential to imagine beforehand the kind of answer that each question will provoke and whether the answer is the one that is required. It is, therefore, a good plan to try out a questionnaire on a few people before having a large number printed. A question that may seem simple and clear cut to the author can quite easily prove a stumbling-block to users of the form. If a trained enumerator is to aid people to fill up the form then the questions asked can be a little more involved than if the survey is to be carried out by post, but it must be remembered that the public is less likely to disclose confidential personal information if the survey is carried out through an enumerator than through an impersonal medium.

2.7 A large amount of statistical data is obtained not by questionnaires but by the investigator himself going out into the field and counting or collecting items. This is the source described earlier as observations. Whatever the type of data it is essential to adopt a system of collection that is both logical and tidy, so that the data will still be understood at some later date. Results should therefore always be recorded in a notebook, since odd scraps of paper are easily lost. The heading of each page must give details of the raw material contained on that page, together with the date and place of collection. It is very tempting to say at the time of collection that this information is so obvious that it need not be recorded, but if the material is not used for some time such an omission may be regretted, especially if a large amount of other material has been used or collected meanwhile. Records should be made with a sharp pencil, as ink is very liable to smudge and become unreadable, especially in outdoor fieldwork. If the data are required for future reference it should be converted to ink later.

2.8 The recording of the data in a notebook is quite straightforward if a systematic method is followed. Suppose a series of barometer readings is being made every quarter of an hour at three different levels in a tall building. Three columns are needed, the first for the time, the second for the level, and the third for the barometric height. The sets of figures would then be placed for ease of recording and reading as in the specimen page shown in fig. 2.4. Anyone looking at the sheet at a future date would be able to understand what had been done and the results would leave no room for ambiguities.

Blackbush Grammar School 15 Oct. 1968
Level A Ground floor Level B Second floor
Level C Fifth floor of main building
Readings of barometer in mm. of mercury

Time	Level	Reading	Time	Level	Reading
10.01	A	767			
10.07	B	761			
10.12	C	759			
10.16	A	766			
10.20	B	763			

Fig. 2.4. Page from observer's notebook

2.9 Imagine that your local town is contemplating the introduction of parking restrictions in the main street as part of a campaign to relieve congestion in the centre of the town. It is desired to have some idea of the amount and nature of the traffic entering the town during the main part of the day. The traffic passing two points during the busy period of the day is therefore counted each day for a week. The necessary observers are stationed in pairs, relieving each other at fairly regular intervals; a specimen page from the notebook of one of the observers is shown in fig. 2.5. It will be noticed that the counting is done by placing the strokes in groups of five, with every fifth stroke put cross-ways to make counting easy. Thus thirteen cars and seven lorries are recorded as having been observed in the period from 10 to 10.15 a.m. Before counting is begun some decisions have to be made to remove any ambiguities that may possibly arise. For example, it must be decided beforehand whether a trader's van is to be counted as a lorry or as a car; and whether a light motor-propelled bicycle should be counted as a motor cycle or omitted altogether.

Date 18 Oct. 1968	Time	
Place Millbridge Rd. G.P.O.	10–10.15 a.m.	10.15–10.30 a.m.
Cars	~~1111~~ ~~1111~~ 111	
Lorries	~~1111~~ 11	
Public Service Vehicles	111	
Motor cycles	1111	

Fig. 2.5. Page from observer's notebook

When the nature of the survey has been agreed the observers can be briefed and then sent out for the actual collection of the data. At frequent intervals the results obtained to date should be collected and inspected to make sure that they conform to the requirements and that no unforeseen snags have arisen. At the end of the week all the data can be collated ready for analysis. At this stage it will consist of large numbers of pages of observers' notebooks, and the next stage in the investigation is to reduce this large volume of figures to a few manageable ones to bring out the particular points of interest. This part of the investigation will be described in the next chapter.

2.10 Examples of circulated questionnaires can be found in reports emanating from the Government Social Survey Unit. Fig. 2.6 (on pages 18 and 19) gives the questionnaire, and covering letter, relating to a survey of motor cyclists. The objective of the survey (carried out in 1960) was to obtain information on the age and experience of owners of motor cycles (including scooters and mopeds), for comparison with independently collected statistics on motor-cycle accidents, in order to estimate the accident rate per mile for different age and experience groups. Some 10,000 questionnaires were sent, out of which about 95 per cent were returned completed.

To obtain this completion rate, a considerable effort had to be put into the follow-up of the initial mailing of the questionnaires. The findings of the survey are described in the report listed at the end of the chapter.

SUGGESTED READING

Some of the following books and pamphlets should be read in order to see the kind of raw material commonly used in statistical work. Most of the items listed will be found in a good public library.

(i) (Her Majesty's Stationery Office routine publications):

Monthly Digest of Statistics.
Board of Trade Journal (weekly).
Financial Statistics (monthly).
Ministry of Labour Gazette (monthly).
Annual Abstract of Statistics.
Economic Trends (monthly).
Abstract of Regional Statistics (annually).
Registrar General's Statistical review of England and Wales (annually).

(ii) (Other publications):

Motor Industry of Great Britain, 1968 (Society of Motor Manufacturers and Traders).
Lessons of the British War Economy by D. V. Chester.
Monthly Bulletin of Statistics (Statistical Office of the United Nations).
Social Surveys by D. Caradog Jones.
London Travel Survey 1949 (London Transport Executive).
Accidents to Young Motor Cyclists, Social Survey Report no. 277B (HMSO).

EXERCISES

The data obtained in many of these exercises will be needed for numerical work in later chapters. As many as possible of the exercises should therefore be carried out. Members of a class could be set a varied assortment and the results collected and retained for future use.

2.1 It is desired to investigate the reading habits of schoolchildren in relation to their age. Design a questionnaire to be used for this purpose remembering that it is important not only to obtain the child's age but

Central Office of Information,
Social Survey Division,
Montagu Mansions,
Baker Street,
LONDON, W.1.

Dear Sir or Madam,

Road Safety

I am writing to ask for your help in an enquiry we are making for the Government Road Research Laboratory.

In carrying out its work on road safety, the Laboratory needs to know who are the chief road users, how much they use their vehicles, and how much experience they have with different kinds of vehicle. At present we are asking only about motor-cyclists, including drivers of motor scooters and mopeds.

We can only get this information by writing to motor-cycle owners. To save expense we are not writing to all, but only to a small number, chosen at random from the registration records. In order to be sure that all points of view are taken into account we are anxious to get a reply from every person we write to.

The questions are on the back of this letter. Would you please fill in the answers and post the sheet back to us, using the enclosed label and envelope. There is no need for a stamp.

Please note that it is *your* reply we want, even if you do not at present ride any machine. Do not ask anyone else to fill in the form instead of you, or we will not have a true cross-section of owners.

Your reply will be kept strictly confidential, and will *only* be used for counting how many people give each different answer. We shall pass these total figures to the road safety authorities, but we shall not mention any names.

I would be most grateful for your help.

Yours faithfully,

C. Scott.

PLEASE TURN OVER

Fig. 2.6. Social survey questionnaire
(Reproduced by permission of Her Majesty's Stationery Office)

1

Vehicle registration No.............................

1. Is the above machine a motor-cycle with side-car? ____________

a motor-cycle without sidecar? ____________

a motor scooter?........ ____________

or a moped or auto-cycle? .... ____________
(A moped or auto-cycle means anything which has a motor and pedals.)

WRITE YES AGAINST ONE OF THESE

2. Are you yourself still the owner of the above machine? ....
(If your answer is NO, please state when you sold it, then answer questions 3 and 4 for the period before you sold it.)

.... .1958
date when sold by you

3. Have you yourself driven the above machine during the last 4 weeks? If so, roughly how far?

(If you did not drive it, write 0) .... miles *in last 4 weeks*

Give rough mileage for the last 4 weeks. You may explain below if that period was very different from your normal.

Please note that the figure you write above should be the mileage you yourself have done, as driver, on the above machine.

4. When did you first take out a PROVISIONAL LICENCE to drive any vehicle?

Month: .... Year: 19.. ....

5. Since that time, how much driving experience have you had?
Please answer below for each type of vehicle.

Motor-cycle *When did you first drive a motor-cycle?* .... 19 ..... WRITE NEVER if
month year never driven.

Have you driven more than 1,000 miles on a motor-cycle? (Write YES or NO)..............

Motor scooter *When did you first drive a motor scooter?* .... 19 ..... WRITE NEVER if
month year never driven.

Have you driven more than 500 miles on a motor scooter? (Write YES or NO)..............

Moped or auto-cycle (with motor and pedals) *When did you first drive a moped or auto-cycle?* .... 19 ..... WRITE NEVER if
month year never driven.

Have you driven more than 250 miles on these? (Write YES or NO)..............

PLEASE DO NOT LEAVE BLANKS. If not sure of the month, give the year only.

6. What is your date of birth? Year: 19...... Month: .... Day: ....

PLEASE RETURN THIS FORM AS SOON AS POSSIBLE USING THE ENVELOPE AND LABEL PROVIDED. PLEASE DO NOT DELAY EVEN IF YOU CANNOT ANSWER ALL THE QUESTIONS

WE WILL BE VERY GRATEFUL FOR YOUR CO-OPERATION

Fig. 2.6. (*cont.*)

also to subdivide the reading matter into the various types of literature. The form can then be duplicated and a survey carried out in the school. To avoid inaccuracies and biased results in the final analysis it is essential to ensure that everybody completes a form.

2.2 A similar survey to that of exercise 2.1 could be carried out to discover how students of various ages spend their spare time.

2.3 Design a questionnaire to be used to find out the method or methods used by students to come to school or college, and the length of time taken.

2.4 A survey is to be carried out amongst schoolchildren to find the number of hours per week that they spend playing various games in relation to their ages. Design a suitable questionnaire for this purpose and then use it in your school to obtain the required data.

2.5 As part of a survey concerning the growth of children it is desired to know the heights, weights and chest measurements of children subdivided according to age. Collect from all the schools in your neighbourhood, as well as from your own, as much data as is possible, remembering to employ a consistent system of recording.

2.6 Make a series of simultaneous readings of the barometer and thermometer every day over a fairly long period of time. The readings should be made at the same time each day. Does there appear to be any form of relationship between the readings of the two instruments?

2.7 Find the amount of space given by a daily newspaper to various types of news such as foreign affairs, parliament, home news, articles, crime, sport and advertisements. Do this over a period of time for a number of different newspapers. (As newspapers vary in size from day to day and devote more space to certain features on some days than others it is best to study each newspaper for a period of a week.)

2.8 During the association football season, collect, for one of the well-known football clubs the attendance at each match together with the result of the match. By repeating this procedure for a number of clubs in the same football league division investigate whether the clubs that have the best match records also have the largest attendances.

2.9 Collect forty conkers from one particular tree and measure (*a*) the maximum length of each conker, and (*b*) the weight of each conker. Repeat the procedure for another tree and see if you can detect any differences between the characteristics of the conkers collected from the two trees.

2.10 Collect a large number of flowering specimens of *Lesser Celandine* and count the number of petals in each flower. The procedure can be carried out first for early flowering plants and then for later flowering plants. On the basis of the figures obtained, can you detect any differences between the two times of flowering?

2.11 Collect a large number (say 200) specimens of a common flower such as a buttercup. Count the number of shoots, leaves and petals on each plant. Repeat the procedure for buttercups collected from a locality as different as possible from the first set and see if you can find any differences between the buttercups of the two localities.

2.12 On a day when fish is being served in the dining hall make arrangements with the superintendent to measure the length of all the fish that are to be cooked. Compare the amount of variation between the lengths of the fish with the average length of the fishes.

2.13 Examine about 300 pods of garden peas and count the number of peas in each pod. Is this number constant or do you find much variation from one pod to another?

2.14 Catch a large number of specimens of a common species of butterfly, say the Cabbage White, and measure the length of the right wing of the butterflies. Do this on a number of occasions over the season and see if you can detect any variation in the wing length over the season. In general terms how would you expect this wing length to vary over the season and why?

2.15 Collect 100 leaves from each of three Great Beech trees and count the veins in the leaves. Do the trees appear to show any difference as far as can be judged by the number of veins in their leaves?

2.16 Children's teeth vary enormously in their soundness. By questioning and examining all the children in the age range 12–16 that are readily available find the numbers of whole teeth and the numbers of 'stopped' teeth in each child. Does there appear to be any difference between boys and girls and between children aged 12 and those aged 16? (The ages chosen could be varified according to the institution concerned.)

2.17 Take a plant for which two types of flower may be distinguished, like a primrose. Collect each week for a period of six weeks fifty primroses of the 'pin-eyed' variety and fifty of the 'thrum-eyed' variety. On each plant count the number of flowers and use the data to determine whether there is any difference in the plants over time or between the two types as judged by the number of flowers.

2.18 In performing experiments in physics designed to determine some physical constant the results obtained vary amongst themselves no matter how carefully the experiment is carried out. To illustrate this perform the usual experiments to determine the specific gravity of a liquid, such as brine, twenty times. Work out the average of the twenty determinations and find the maximum amount by which any single determination differs from that average. This shows the sort of error that might occur if the result was based on a single determination.

2.19 Repeat the same procedure as in the previous exercise but this time measure the coefficient of expansion of a metal such as copper twenty times and find the maximum variation in any one determination from the average.

2.20 The focal length of a convex lens may be obtained using a pin as an object and fixing the image with another pin by means of the method of parallax. Using the object and image distances the focal length can now be found from the usual formula $\frac{1}{f} = \frac{1}{u} + \frac{1}{v}$. Perform this experiment twenty times with the same lens using a slightly different value of object distance (u) each time. Work out the focal length from each experiment and see how much variation occurs in the focal length from one experiment to another.

2.21 Telephone calls vary greatly in length. Select a railway station or post office where there are several kiosks in a row and keep a record over a period of an hour of the times that people enter and leave the kiosks. Hence obtain a series of telephone call lengths for the period. From this determine the proportion of calls that are under 5 min., between 5 and 10 min., or over 10 min. Repeat the whole experiment in another locality and see if these three proportions differ.

2.22 When a fruit tree, such as an apple tree, is stripped, weigh each item of fruit before putting it in the collecting basket. Work out the average weight of the apples and also find a weight above which 90 per cent of the apples lie. Repeat the experiment for an apple tree of a different variety and see if either the average weight or the weight above which 90 per cent of the apples lie is different.

2.23 Perform the following simple experiment fifty times. A line is drawn of 12 in. in length. Preferably each line should be on a separate sheet of paper. Then by eye make a mark 3 in. from one end. After fifty attempts measure the actual distance that has been cut off by eye each time. Now perform the whole experiment a second time, only on this occasion measure the portion cut off as soon as each individual mark has been made. Repeat for all the fifty lines. Does the second method appear to give more accurate results than the first method?

2.24 Open a telephone directory at a page which does not contain any advertisements. Count how many telephone numbers in a column end with the digit 1. Repeat this procedure for about ten different columns and see if the proportion of numbers ending in 1 differs at all.

Next repeat the whole procedure, this time counting the number of

9's in the last digit. Do you find that there are more or fewer 9's than 1's in the last digit and can you account for any difference you may find?

2.25 Take a novel by some well-known author and selecting a dozen pages count the number of words in each sentence. Repeat the procedure for a number of different authors and types of book. Can you notice any differences between the styles of the authors that are brought out by this investigation?

2.26 From a newspaper obtain the prices of a number of types of shares each week over a period of some months. Take one type of stock or share from each of the following groups: government stock, newspaper company, oil company, textile company, tin company, shipping company, mining company. From these figures see whether the prices all go up and down together or whether some go up more or less than others. Try to explain these rises and falls in relation to current world events.

2.27 Find a zebra crossing in a town. Over a period of, say 2 hr. make a count of the number of persons using the crossing and also the number of persons crossing the road for 50 yards on either side of the zebra crossing. It is best to carry out this exercise using at least six observers and to make one pair responsible for the actual crossing and one pair responsible for the 50-yard stretch on either side of the crossing.

2.28 Repeat exercise 2.27, but this time subdivide all persons into men, women and children, making certain that there is a firm definition to distinguish adults from children and that this definition is the same for all the observers.

2.29 Take two different points on a road, one just past a traffic light or point-duty policeman and the other well away from either. At each point count the number of motor-vehicles passing in successive intervals of 15 sec. This should be done for an hour and repeated at different times of the day. It will be essential to have at least two observers, one of whom is responsible for the time-keeping. If possible there should be two pairs of observers to act as reliefs. Are the variations that occur from interval to interval different for the two points and for different times of the day?

2.30 Repeat exercise 2.29 dividing the traffic up into various classes such as cars, lorries and so forth. A comparison could also be made of various roads in the neighbourhood.

3

THE TABULATION OF DATA

3.1 In the previous chapter the investigations made led to the collection of a large amount of raw materials in the form of entries in an observer's or experimenter's notebook. From this mass of figures it is usually difficult for anyone to pick out the salient features of the data without missing important points. Table 3.7 later in this chapter gives the heights of a hundred 14-year-old schoolboys, and it is difficult to pick out very much from the table except that the heights seem to range from about 50 in. to just below 70 in. For this reason it is essential to condense the raw figures into some more manageable form which will enable the investigator to pick out at once those features that he wishes to pursue further. A first step is to compile a carefully designed table in which figures possessing similar properties are grouped together.

Table 3.1. *Numbers of vehicles on the Great North Road*

Motor cars	Lorries	Public service vehicles	Motor cycles	Total
412	273	64	39	788

Suppose an investigation has been made into the number of motor-vehicles proceeding north along the Great North Road in a period of 2 hr. past a certain point. The raw material has been collected by posting a pair of observers, and a record kept in the notebook in the manner described in chapter 2. At this stage the notebook consists merely of a mass of figures and strokes in the various categories. It is a fairly straightforward matter to count up the strokes, which will be in groups of five if the method suggested earlier is being used, and record the results as in table 3.1. It will be noticed that the largest figure is on the left and the smallest is on the right. This is customary and makes it easier to grasp the purport of the table, especially if there are large numbers of categories. Sometimes all the figures in the table are reduced to percentages of the total. It is much easier, and conveys a more vivid impression, to say that 52·3 per cent of the vehicles, rather than 412 out of 788 vehicles, were cars. Thus table 3.1 would be replaced by table 3.2.

Table 3.2. *Percentage distribution of 788 vehicles on the Great North Road*

Motor cars	Lorries	Public service vehicles	Motor cycles	Total
52·3	34·6	8·1	5·0	100·0

One rule to be strictly observed in such tables is that the total number of observations on which the percentages are based, in this case 788, must be given.

3.2 In a count of vehicles, each observation goes into one of a number of categories. As another example of a classification into categories, each adult in a town might be recorded as to his or her marital status, that is whether single, married, widowed or divorced. In both these cases the observation does not involve anything directly measurable, but merely the placing of each observation into one of a number of descriptive categories. The observations in this case are said to be **qualitative**, but it must not be thought that it is necessarily impossible to place the categories in an appropriate order. For example, a headmaster was asked to place his pupils into one of four categories according to their general appearance at school. The categories were: well-dressed, average, below average, and very shabby. The numbers were as in table 3.3.

Table 3.3. *Standard of dress of pupils*

Well-dressed	Average	Below average	Very shabby	Total
21	73	39	16	149

The order in which the categories have been placed is the logical one and implies that the standard of dress decreases as the table goes from left to right. To put the categories in order of size would produce a table whose true meaning was not obvious without careful study.

3.3 The next case to be considered, is where the observation is **quantitative** and consists of a measurable quantity, such as height or weight or temperature, and is more common. Sometimes the observation can take integral whole numbers only. As an example, table 3.4 gives the number of calls made to a local fire station each day for a year, the order of the days being across each row. Thus on the first day there was one call, whilst on the second day there were no calls and so on.

Table 3.4. *Calls to local fire station per day for one year*

1	0	1	2	0	0	3	0	2	1	1	0	4	2	1	0	1	2	0	1
2	5	0	1	3	0	1	2	0	2	1	0	0	0	4	0	3	1	0	3
0	1	3	0	2	0	0	1	1	4	2	0	4	2	0	2	0	4	2	0
2	0	1	0	4	1	2	0	0	2	1	3	0	2	1	2	3	0	0	3
3	2	0	3	2	0	1	2	0	2	4	0	1	2	1	0	1	2	0	4
2	0	2	1	0	4	0	3	1	1	2	0	4	2	0	3	1	2	4	0
3	4	0	3	1	0	0	2	1	2	0	2	0	3	1	1	0	0	0	4
0	1	2	0	0	1	1	0	1	5	0	2	3	2	0	0	4	1	2	0
2	0	1	3	2	0	0	3	1	0	1	3	0	1	0	2	0	1	3	2
3	0	0	0	1	2	1	0	1	2	2	1	1	0	1	2	1	0	0	0
2	0	1	2	1	0	0	1	1	5	0	1	0	3	1	0	4	0	1	0
1	1	0	2	0	1	2	0	0	1	3	0	0	1	1	0	1	2	0	1
0	4	2	0	0	0	1	1	0	1	1	2	3	0	1	1	3	0	4	0
1	0	1	5	0	1	1	2	3	0	0	1	1	0	0	0	1	1	3	0
1	2	0	3	1	0	2	1	0	1	3	0	0	3	2	1	0	0	1	0
1	0	2	1	0	1	3	0	3	1	1	1	2	0	1	0	1	5	0	1
1	2	0	1	3	2	0	3	1	0	1	4	2	1	0	3	1	2	1	0
1	0	3	2	1	1	0	1	4	2	0	1	3	0	1	2	3	0	1	1
1	2	1	0	1															

To reduce such a set of data the first step is to group the figures together so that all the days with the same number of calls are in one group. This is illustrated in table 3.5, where the strokes for each of the first 40 days have been placed in the appropriate groups. The procedure is continued for all the remaining days so that finally the table contains a stroke for each day on the appropriate line. These strokes are now counted up and recorded as in table 3.6. The table shows that on 130 days in the year no calls were made,

Table 3.5. *Grouping of calls*

No. of calls	No. of days
0	~~1111~~ ~~1111~~ ~~1111~~
1	~~1111~~ ~~1111~~ 1
2	~~1111~~ 11
3	1111
4	11
5	1

Table 3.6. *Number of calls to fire station each day for a year*

No. of calls	No. of days with that number of calls
0	130
1	109
2	64
3	38
4	19
5	5
Total	365

on 109 days one call was made, on 64 days two calls were made and so on, until finally on 5 days in the year no less than five calls were made. On no day were there more than five calls. The table is wholly free from ambiguity and there is no difficulty in distinguishing between, say, the case of no calls and the case of one call. It was virtually impossible to see the salient features from the original data, such as the fact that on about one day in three there are no calls, but this becomes obvious from table 3.6. Such a table is called a *frequency distribution*. Once again, the numbers involved in the table can be expressed as percentages remembering to give the total, in this case 365, on which the percentages are based.

The total number of calls made in a year could be found quite simply from table 3.6. On 5 days there were five calls, giving a total of 5×5 calls for those 5 days. On 19 days there were four calls, a total of 19×4 calls, and so on. Hence the overall total number of calls is

$$5 \times 5 + 19 \times 4 + 38 \times 3 + 64 \times 2 + 109 \times 1 + 130 \times 0 = 452.$$

Thus 452 calls were made in the year and this is also the sum of the original individual entries in table 3.4.

3.4 More usually the observations do not take integral values only, but may take any numerical value subject to the limitations of the apparatus used for measuring. Suppose, for example, that the heights of 100 schoolboys aged 14 years have been measured, and the observations recorded in a notebook in columns of figures, as in table 3.7. A scale marked in tenths of an inch has been used, and the height judged to the nearest mark. This implies that it is correct to the nearest tenth of an inch. For instance if a boy's height is between 62·3 and 62·4 in. then it is recorded as 62·3 in. if it appears nearer 62·3 than 62·4 in., otherwise it is put down as 62·4 in. The first step in forming a table is to pick out from the 100 measurements the smallest and the greatest heights. These are found to be 54·3 and 68·3 in. respectively, and are italicised in the table. Now it is usual to have not more than about fifteen groups in the final table, which can be achieved in this case by having a series of groups each containing heights within a range of 1 in. The procedure to be followed is shown in table 3.8. Each group is defined in the left-hand column and the number of schoolboys whose height falls in each group is recorded in the right-hand column by means of a stroke. The groups are defined in such a way that there is no possibliity of any ambiguity arising. For instance, the first boy has a height of 63·3 in. and thus goes into the group that is labelled '63·0–63·9 in.'. The second boy has a height of 60·0 in. and goes into the group labelled '60·0–60·9 in.'. This procedure is carried out for each of the 100 boys. The rough grouping has already brought out the fact that the bulk of the heights are clustered around about 62 in., a fact which would be much more difficult to spot from the unsorted heights in the notebook. Using the rough table just formed, a final table can be made (table 3.9). In this table it is important to note the way in which the groups are defined.

The symbol 54– indicates that all boys with heights of at least 54 in. but under 55 in., that being the lower limit of the next group, will go into this group. Similarly 55– will mean that all boys with heights of at least 55 in. but under 56 in., the lower limit of the next group, will go into that group. At this point, however, a little care must be exercised. Anyone looking at table 3.9 would

Table 3.7. *Heights of schoolboys in inches*

63·3	58·1	63·8	61·9	65·4	62·3	61·4	66·9	63·0	60·1
60·0	64·6	61·1	62·9	60·6	59·3	65·2	61·2	57·8	62·4
65·9	62·0	66·1	58·9	64·1	64·2	61·3	60·8	67·0	65·0
63·1	63·7	61·8	64·3	62·6	*54·3*	63·2	63·4	61·8	64·7
59·2	62·3	60·7	65·6	61·1	63·8	60·5	62·8	64·1	61·9
64·0	61·7	64·5	61·3	60·2	61·2	66·3	59·4	*68·3*	62·3
66·7	65·7	62·8	64·9	62·7	62·2	61·9	62·6	63·9	64·5
62·4	67·9	63·4	55·7	63·1	59·7	64·8	65·8	60·4	62·8
61·7	62·4	61·9	63·8	61·6	62·1	63·0	64·4	62·0	61·5
62·5	63·2	62·1	62·8	63·4	62·9	60·3	62·1	61·3	63·9

Table 3.8. *Grouping up of data*

Group of heights (in.)	Heights in group				
54·0–54·9	1				
55·0–55·9	1				
56·0–56·9	—				
57·0–57·9	1				
58·0–58·9	11				
59·0–59·9	1111				
60·0–60·9	~~1111~~	1111			
61·0–61·9	~~1111~~	~~1111~~	~~1111~~	111	
62·0–62·9	~~1111~~	~~1111~~	~~1111~~	~~1111~~	11
63·0–63·9	~~1111~~	~~1111~~	~~1111~~	1	
64·0–64·9	~~1111~~	~~1111~~	11		
65·0–65·9	~~1111~~	11			
66·0–66·9	1111				
67·0–67·9	11				
68·0–68·9	1				

Table 3.9. *Height of schoolboys*

Height (in.)	No. of schoolboys	Height (in.)	No. of schoolboys
54–	1	62–	22
55–	1	63–	16
56–	—	64–	12
57–	1	65–	7
58–	2	66–	4
59–	4	67–	2
60–	9	68–	1
61–	18	Total	100

assume that all boys with heights from exactly 54 to exactly 55 in. would go into the first group. On referring back to table 3.8 it can be seen that all boys whose heights are 54·0, 54·1, 54·2, ..., 54·9 in. have been put into the first group. Further, it has already been demonstrated that a boy whose height is recorded as 54·0 in. may have an actual height that is anything from half-way between 53·9 and 54·0 in. to half-way between 54·0 and 54·1 in., that is from 53·95 to 54·05 in. Similarly, a boy whose height is recorded as 54·9 in. may in fact have a height that is anything from 54·85 to 54·95 in. Hence the actual range of heights in the group described in the table as 54– are from 53·95 to 54·95 in., since any boy whose height falls between those two limits will be included in this group. It is clear, then, that the method of defining the groups in table 3.9 is not perfect and it can be improved upon in one of the three ways given in the next two sections.

3.5 *Method* (i) In the first place table 3.9 may be modified quite simply by writing into the table the exact boundaries of each of the groups. These boundaries give the group limits as 53·95–54·95 in., 54·95–55·95 in. and so on. No measured height in table 3.7 falls on one of these boundary values and it is thus quite clear into which group each individual falls.

Method (ii) Instead of giving the values of the boundaries of the groups the central or middle value of each group is given. If a group contains values from 53·95 to 54·95 in. then the central value is half-way between these values, that is at $\frac{1}{2}(53{\cdot}95+54{\cdot}95)$ or 54·45 in. Similarly the central value of the next group is 55·45 in. Table 3.9 would then become table 3.10.

Table 3.10. *Heights of schoolboys*

Height (in.), central values	No. of schoolboys	Height (in.), central values	No. of schoolboys
54·45	1	62·45	22
55·45	1	63·45	16
56·45	—	64·45	12
57·45	1	65·45	7
58·45	2	66·45	4
59·45	4	67·45	2
60·45	9	68·45	1
61·45	18	Total	100

The group interval is defined as the range of values in any one group. In this case it is 1 in. and may easily be obtained by subtracting one central value from the next one, that is 55·45 – 54·45 or 56·45 – 55·45 and so on. It is also equal to the difference between the upper and lower limits of any one group. For instance, the first group has 53·95 as the lower limit and 54·95 as the upper limit, and these values differ by an inch.

To obtain the group boundaries from a table that gives only the group central values it is necessary to take: central value of group + half the group interval, and central value of group – half the group interval. For example, the group with central value 62·45 has as its group boundaries 62·45 + 0·5 and 62·45 – 0·5, i.e. 62·95 and 61·95 in.

3.6 *Method* (iii) There is one other method of dealing with the grouping that will be described, although it will not be used in this book. The methods of section 3.5 result in tables that are technically correct but not very pleasing to the eye owing to the rather awkward central value. One possible way to tidy up the table would be to take the group boundaries at exactly 54 in., exactly 55 in., and so on. This is a feasible solution, but a number of observations will fall exactly on the group boundaries and there must be some rule to determine how these are to be tabulated. Two suggested rules are:

(*a*) Put the first observation that occurs on a group boundary in the higher of the two possible groups, the next observation in the lower of the two groups, the next in the higher and so on. For instance, suppose that the first two heights were 56·0 and 59·0 in. respectively. Then 56·0 would be put in the group 56–57 in. whilst 59·0 would be put in the group 58–59 in. This procedure would be adopted throughout the observations.

(*b*) Put a half observation in each of the groups on either side of the observation. If the first two heights were as in (*a*) then for 56·0 a half would be put in each of the groups 55–56 in. and 56–57 in., whilst for 59·0 a half would be put in each of the groups 58–59 in. and 59–60 in. This may seem a rather artificial method but, provided that there are a reasonable number of observations, it should result in little or no error in subsequent calculations based on the table.

The second method (*b*) will be adopted for illustration. Repeating the procedure outlined in section 3.4, and using this rule for boundary cases, table 3.11 is obtained. It will be seen that although the artificial concept of half a boy has been introduced, a gain has been made, in that the groups have convenient boundaries and the central values are now 54·5, 55·5 and so on. Frequently the variable will be measured to an accuracy sufficient to avoid all this trouble in the groupings. For example, has all the heights been measured to two places of decimals and the groupings of table 3.11 adopted, only heights such as 64·00 or 59·00 in. would fall on the boundaries, and hence have to be divided between the upper and lower groups. If the heights are measured to two places of decimals, it is unlikely that more than about one in a hundred of the observations will fall exactly on a boundary. This is an extremely small proportion and the greater the accuracy of measurement the less trouble there will be in forming the groups for a table. Nevertheless since the method is not unambiguous and easily leads to mistakes, it is not recommended for general use.

Table 3.11. *Heights of schoolboys*

Height (in.)	No. of schoolboys	Height (in.)	No. of schoolboys
54–55	1	62–63	22
55–56	1	63–64	15·5
56–57	—	64–65	12
57–58	1	65–66	6·5
58–59	2	66–67	4·5
59–60	4·5	67–68	1·5
60–61	8·5	68–69	1
61–62	19	Total	100

3.7 The tables in the previous sections have all used the same grouping interval over the whole range of the variable, the variable in this case being height. The chief reason for using equal class intervals is that the numbers in each interval, or frequencies as they are called, are then directly comparable. Sometimes, however, the observations are very close together at one end of the distribution but are far apart at the other end, so that, if a suitable group interval is used for the lower end, there will be large numbers of groups at the other end with few or no observations in them.

Table 3.12. *Deaths of males and females under 35 years of age in England and Wales, 1966*

Age at death (years)	No. of deaths
0–5	18,930
5–10	1,341
10–15	1,104
15–20	2,738
20–25	2,467
25–30	2,298
30–35	2,876
Total	31,754

Consider the data in table 3.12 relating to the deaths of males and females in England and Wales in 1966. The group intervals are all equal and from the table one might deduce that whereas in the early 'teens about 221 deaths a year occur (that is, one-fifth of 1,104) the deaths in the first 5 years of life were some 3,800 per year (that is, one-fifth of 18,930). The first part of this statement is correct, the second part erroneous. A closer inspection of the figures from which this table was constructed reveals that the deaths are more or less evenly spread over all the class intervals, with the exception of the first, where the vast majority of deaths occur under 1 year of age. To illustrate this, the first age-group in table 3.12 has been further subdivided in table 3.13 to illustrate how the deaths are in fact spread over those first 5 years. From a study of this table it becomes clear that nearly one third of the deaths occurring under 5 years of age occur during the first 4 weeks of life. This is obscured in the original table 3.12.

Table 3.13. *Deaths under 5 years subdivided*

Age at death	No. of deaths
Under 4 weeks	5,369
4 weeks–3 months	2,098
3–6 months	1,737
6–21 months	1,379
1–2 years	1,124
2–3 years	682
3–4 years	515
4–5 years	462
Total	18,930

3.8 Suppose now that the two tables just considered were combined in order to bring out how the deaths were divided amongst the age groups. This would produce a composite table with varying group intervals and would make comparisons between age-groups somewhat difficult. To obviate this difficulty a further column could be inserted giving the number of deaths for some fixed interval. A period, such as a year, is chosen, and the number of deaths that would occur in a year calculated, assuming that the rate of occurrence of deaths inside the group is constant. It does not matter what length of interval is selected provided it is retained throughout, but the use of such a fixed interval enables comparisons of the rate of deaths for various age-groups to be made easily. The figure of 69,797 does not mean that 69,797 deaths took place, but that if the rate of deaths for the first 4 weeks of life was continued throughout the first year, the resulting number of deaths would reach that rather large figure.

Table 3.14. *Composite table of deaths*

Age at death	No. of deaths	Deaths per year	Age at death	No. of deaths	Deaths per year
Under 4 weeks	5,369	69,797	4–5 years	462	462
4 weeks–3 months	2,098	13,637	5–10 years	1,341	268
			10–15 years	1,104	221
3–6 months	1,737	6,948	15–20 years	1,366	273
6–12 months	1,379	2,758	20–25 years	2,467	493
1–2 years	1,124	1,124	25–30 years	2,298	460
2–3 years	682	682	30–35 years	2,876	575
3–4 years	515	515			

3.9 All the preceding work has been concerned with just one variable measurement for each observation, but often there is more than one possible measurement for an individual. Suppose that each individual has had two characteristics measured. Thus the schoolboys of section 3.4 may have had their weights, as well as their heights, recorded, and in table 3.15 these weights are given with the boys arranged in the same order as for the heights recorded in table 3.7. To illustrate how these two characteristics vary with each other a joint table is formed. To make this table, a framework is first drawn up and each observation of height and weight put into its appropriate cell. The framework is shown as table 3.16, and each observation is represented by a stroke. Thus the first observation has height 63·3 and weight 123·1, so that it goes in the column headed 62 and in the row labelled 120. This procedure can be repeated for the 100 observations. If any observation falls exactly on a group boundary a half is put in the cells that fall either side of the boundary. The table, when completed, can have its rows and columns added up as in table 3.17. The right-hand column gives the total numbers appearing in each row of the table and would be the same result as if a table for weight alone had been compiled. Similarly the bottom row gives the total number in each column of the table and is effectively a distribution of height alone. The sums of these two marginal distributions should each equal the total number of observations.

Table 3.15. *Weights of schoolboys (lb.)*

123·1	131·7	129·6	120·8	127·0	125·4	112·9	121·3	135·8	128·9
117·5	131·2	131·1	123·4	121·8	121·8	131·3	123·7	117·2	121·2
131·7	125·1	127·2	126·0	131·9	134·5	119·2	126·6	131·2	127·7
127·4	127·4	126·0	125·1	127·3	112·8	125·7	127·2	115·4	131·6
125·3	122·5	120·9	127·2	125·8	121·4	117·1	120·8	129·3	121·0
127·2	123·8	129·7	120·6	125·5	130·7	124·6	122·1	134·1	121·5
131·9	131·1	125·3	131·7	130·2	129·6	121·8	131·4	116·5	133·0
125·8	134·0	121·8	120·3	117·4	122·8	131·7	129·6	120·5	131·4
121·6	125·9	114·5	133·2	115·9	130·3	127·4	125·8	120·7	126·6
126·3	125·8	122·4	127·2	123·7	121·9	122·6	126·5	123·0	127·1

Table 3.16. *Observations of height and weight*

Weight (lb.)	Height (in.) 54–	56–	58–	60–	62–	64–	66–	68–
112–	1			1111				
116–		1		111	11			
120–	1		111	~~1111~~ ~~1111~~ 11	~~1111~~ ~~1111~~ 11		1	
124–			11	~~1111~~	~~1111~~ ~~1111~~ ~~1111~~ 1	~~1111~~ 1	11	
128–			1	1111	~~1111~~ 1	~~1111~~ ~~1111~~ 1	11	
132–					11	11	1	1

Such tables are called *bivariate tables* and are useful in bringing out the relationship between two variables. From the original table it is impossible to see whether the height is related to the weight of the schoolboy or not, but table 3.17 shows that, in general, the taller the boy the greater his weight.

Table 3.17. *Two-way table of height and weight*

Weight (lb.)	Height (in.) 54–	56–	58–	60–	62–	64–	66–	68–	Totals
112–	1	—	—	4	—	—	—	—	5
116–	—	1	—	3	2	—	—	—	6
120–	1	—	3	12	12	—	1	—	29
124–	—	—	2	5	16	6	2	—	31
128–	—	—	1	3	6	11	2	—	23
132–	—	—	—	—	2	2	1	1	6
Totals	2	1	6	27	38	19	6	1	100

Cases will arise where a bivariate table is to be formed from data for which there is no obvious numerical scale for one, or even for both, of the variables. The principles adopted are precisely the same as for one variable only. Thus the data in table 3.18 are classified first by age, which is on a numerical scale, and secondly by sex, which does not have a numerical scale but is simply one or other of two alternative categories. One of the variables, age, is thus quantitative, and the other, sex, is qualitative. As before, any quantitative scale has the groups arranged in order of the quantity, but for a qualitative scale the order if often the order of magnitude unless there is some other natural order for the groups.

Table 3.18. *Full-time students entering universities at undergraduate level for a degree, Great Britain, October 1965*

	Age on entering University: Under 18	18 but under 19	19 or over	Total
Men	870	16,096	21,193	38,159
Women	574	8,142	6,766	15,482
Total	1,444	24,238	27,959	53,641

3.10 The information collected by the observer or available to him has now been reduced from a long series of figures in a notebook into a few well chosen and appropriate tables. Sometimes the tables themselves are sufficient to prove the point under discussion and no further analysis is necessary. But in many cases the tabulation has merely put the raw data in a form suitable for further comparisons.

Frequently the tables produced contain so much information that it is still impossible to disentangle the required bits and pieces of information. In these circumstances a further series of smaller tables is needed. Alternatively it may be possible to draw some form of diagram, which many people will find easier to understand than a table of figures, to present in a clear and forceful manner, the important aspects of the information.

EXERCISES

Much of the data that has been obtained in the exercises at the end of chapter 2 may now be tabulated. The questions asked will probably be answered very much more easily from the tables than from the raw data.

3.1 The number of telephone calls received at an exchange in 120 successive intervals each of ½ min. duration are given below. Form a frequency distribution of the figures.

6	3	4	6	5	1	2	7	2	4	4	4	4	6	4
4	6	5	7	1	1	5	0	0	3	3	5	5	3	4
6	4	6	2	5	5	3	3	6	4	4	2	3	3	3
4	2	1	6	4	6	3	0	3	3	4	4	4	5	3
5	2	7	6	2	6	5	2	4	3	4	3	7	4	4
7	3	5	5	4	5	5	3	1	5	6	1	4	6	3
3	2	0	5	1	6	5	6	1	3	1	4	1	5	6
6	5	6	1	6	7	5	6	3	6	1	3	7	2	5

3.2 The number of words per sentence in eighty sentences chosen at random from *English Saga* by A. Bryant (excluding conversation and quotations) were as follows:

14	28	36	55	22	27	10	22	19	48
28	8	24	18	18	31	32	17	14	41
25	17	22	18	31	38	11	17	42	29
13	83	2	29	13	41	21	71	9	30
26	13	22	37	47	6	25	15	47	35
24	42	12	10	29	18	33	3	31	19
16	7	26	33	19	67	53	12	13	52
8	60	19	31	15	21	8	47	38	25

Draw up a frequency table for this data. To make it reasonably compact use groups of 1–9, 10–18, 19–27 words and so on, and thus have about ten groups. If this is not done a multiplicity of groups with small numbers will result.

3.3 The following table gives the strength, in pounds, of sixty samples of cement mortar. Form a table, having about eleven groups, from the observations. Notice that all the results are whole numbers and hence the table could be formed (*a*) by making the groups go from, say, 490–500, 500–510, etc., or (*b*) by going from 489·5–499·5, 499·5–509·5, etc. Construct tables using both methods.

536	492	528	572	582	506	544	502	548	562
534	542	570	578	532	562	524	548	530	592
564	536	540	530	590	554	530	560	572	526
542	556	590	546	564	522	570	540	546	532
580	556	574	536	558	570	540	546	560	544
576	490	572	578	586	550	540	542	546	570

3.4 The following figures give the weights of fifty pigs used for a feeding trial. The weights are in pounds.

195	177	180	200	197	170	150	180	192	184
194	204	200	201	195	187	191	200	208	218
203	190	221	173	185	225	190	201	174	180
226	230	218	235	197	210	217	205	200	191
242	225	205	228	196	196	230	170	216	175

(*a*) Form a frequency table of this data using eight groups.

(*b*) If you were told that the ten pigs in each row of the table came from the same litter would you consider that there were any differences between the litters?

3.5 The following data give the yield in pounds of roots of mangolds in forty equal sized plots. Form the data into a table having about eight groups of equal width.

339	384	322	331	277	299	332	302	318	306
314	325	330	328	301	316	338	310	304	302
300	329	322	310	309	342	350	335	320	278
369	341	344	324	316	342	351	324	310	309

3.6 The following figures give the estimated diameters of forty-two nylon threads in units of thousandths of an inch.

1·10	0·96	0·86	1·02	1·08	0·92	0·88	1·06	1·04	0·88
0·92	1·10	1·16	1·06	1·10	1·00	1·04	1·04	1·06	0·98
1·06	1·10	1·00	1·06	0·98	1·04	1·06	1·08	1·00	0·92
0·86	1·02	0·90	1·04	0·92	0·88	1·06	0·92	1·02	1·04
1·04	1·16								

Form a frequency table from these figures with about six groups of equal width.

3.7 The ages in years at which tuberculosis was first observed in 125 machine printers were recorded. Make a frequency table of this data (adapted from data of A. Bradford Hill) using about twelve groups.

16·7	44·7	30·7	30·3	35·7	59·5	53·9	46·1	42·9	32·1
32·9	37·1	55·7	42·5	59·1	38·7	43·0	31·0	37·3	42·5
54·7	48·4	57·7	46·7	26·7	44·5	38·2	24·7	19·9	57·8
59·3	17·6	42·5	26·6	34·5	64·7	36·3	54·0	53·6	49·0
30·4	34·4	55·6	34·1	33·9	59·0	52·7	57·8	48·4	36·1
32·5	47·4	32·4	31·2	45·5	58·1	50·8	50·0	43·4	48·0
43·8	45·2	38·8	16·1	29·5	41·1	48·3	40·0	40·4	49·3
32·7	16·1	57·2	46·8	48·1	47·0	49·3	47·7	48·7	45·0
28·8	58·8	55·7	49·5	41·6	40·3	45·6	41·0	42·8	55·5
27·4	34·6	33·6	49·2	35·3	43·4	59·0	33·1	44·2	41·3
55·0	48·0	46·7	47·2	38·3	38·9	33·0	31·7	21·5	23·2
28·6	53·5	50·9	45·6	48·7	41·2	50·4	39·3	35·8	30·2
46·9	43·3	52·5	59·2	51·6					

3.8 The following data give the percentage ash content in 100 wagons of coal. Form a frequency table with about ten groups, selecting boundaries for the groups such that no observations fall on the boundary.

18·3	12·2	16·8	21·0	17·9	20·2	14·3	17·8	17·0	14·9
10·4	16·3	16·5	18·0	11·9	15·8	15·4	18·6	17·0	12·4
16·4	14·4	16·0	14·6	20·8	17·0	17·1	18·4	16·0	17·5
18·0	18·5	18·3	16·4	16·9	15·1	19·0	18·4	18·5	17·5
19·2	16·4	17·9	18·5	15·1	17·2	15·1	19·8	16·5	14·0
19·3	12·8	18·8	18·8	19·3	16·6	18·1	17·2	17·6	20·3
14·8	16·5	17·2	19·3	20·0	19·1	15·0	20·2	18·1	13·2
15·0	19·3	20·0	16·5	17·6	18·3	18·8	18·1	16·1	21·3
20·0	15·9	17·1	15·6	15·4	18·1	15·5	14·7	15·8	16·7
14·9	20·7	18·5	20·9	17·3	17·4	22·2	15·7	16·8	14·8

3.9 The table below gives the average wholesale price of butter per hundredweight in Chicago over a period of thirty-six months. Plot a graph of this data, and attempt to estimate whether or not the price varies according to the season of the year.

	Jan.	Feb.	Mar.	Apr.	May	June
1931	28·5	28·4	28·9	26·1	23·7	23·3
1932	23·6	22·5	22·6	20·1	18·8	17·0
1933	18·8	18·7	18·2	20·7	22·5	22·8

	July	Aug.	Sept.	Oct.	Nov.	Dec.
1931	24·9	28·1	32·5	33·8	30·9	30·6
1932	18·2	20·3	20·8	20·7	23·3	24·1
1933	24·5	21·3	23·6	24·0	23·6	20·1

3.10 Draw up (*a*) a table for tensile strength, (*b*) a table for hardness, and (*c*) a two-way table for tensile strength and hardness, using the data below which give the tensile strength, in units of 1,000 lb. per sq. in., and the hardness, measured by Rockwell's coefficient, of sixty test pieces of a certain aluminium die-casting (data due to W. A. Shewart).

Sample no.	Tensile strength	Hard-ness	Sample no.	Tensile strength	Hard-ness	Sample no.	Tensile strength	Hard-ness
1	29·3	53·0	21	25·8	69·1	41	29·7	80·4
2	34·9	70·2	22	23·7	53·5	42	32·6	76·7
3	36·8	84·3	23	28·7	64·3	43	32·8	82·9
4	30·1	55·3	24	32·4	82·7	44	30·4	55·0
5	34·0	78·5	25	28·2	56·7	45	38·6	83·2
6	30·8	63·5	26	34·0	70·5	46	28·2	62·6
7	35·4	71·4	27	34·5	87·5	47	29·2	78·0
8	31·3	53·4	28	29·2	50·7	48	35·6	84·6
9	32·2	82·5	29	28·7	72·3	49	34·3	64·0
10	33·4	67·3	30	29·8	59·5	50	34·8	75·3
11	37·7	69·5	31	29·3	71·3	51	40·6	84·8
12	34·9	73·0	32	28·0	52·7	52	28·9	49·4
13	24·7	55·7	33	31·9	76·5	53	34·6	74·2
14	34·8	85·8	34	27·6	63·7	54	31·2	59·8
15	38·0	95·4	35	31·7	69·2	55	33·8	75·2
16	25·7	51·1	36	30·8	69·2	56	34·9	57·7
17	25·8	74·4	37	32·0	61·4	57	36·7	79·3
18	26·5	54·1	38	36·6	83·7	58	32·3	67·6
19	28·1	77·8	39	41·6	94·7	59	34·4	77·0
20	24·6	52·4	40	30·5	70·2	60	34·7	74·8

3.11 The following table gives the weekly number of injuries per thousand employees for the Tenlex Company over a year. Form a frequency table with about ten groups and comment on your table.

3·0	3·9	3·9	2·9	4·6	3·4
3·2	2·9	4·4	3·5	3·3	3·4
3·7	3·8	3·2	3·1	4·3	3·5
3·7	3·3	3·8	3·1	2·1	3·7
2·8	3·1	4·7	2·7	3·5	3·6
3·7	1·8	2·3	2·8	4·0	3·5
3·3	5·3	4·3	3·4	3·6	2·6
4·0	2·6	4·2	2·4	3·2	
3·2	4·0	3·0	3·3	2·5	

4

THE PICTORIAL REPRESENTATION OF DATA

4.1 The last chapter has shown how tables can facilitate the reduction of the observer's raw data and material to a form which enables the reader to grasp the essential features portrayed. In this chapter a further stage in this reduction is dealt with in the construction of charts and diagrams, which enable the salient features of a set of data to be picked out and vividly portrayed so that the reader can spot, without detailed study of the individual figures, the features of particular interest. The primary consideration to be borne in mind in the construction of any chart or diagram is clarity, since a confused diagram is of little help and it is probably better to have no diagram at all, than one that is virtually impossible to understand without a great deal of effort on the part of the viewer. To achieve this standard it is essential to decide at the outset on the purpose of the diagram and to exclude all irrelevant matter from consideration.

Broadly speaking, different considerations are involved according to whether the data are concerned with qualitative or quantitative characters. In the former case the study is of some characteristic such as hair colouring, for which it is difficult to have a numerical scale, whereas for quantitative characters, such as the height of schoolboys, it is possible to have a continuous numerical scale whose accuracy is limited only by the inability of the measuring apparatus to record heights to an accuracy of less than about, say $\frac{1}{8}$ in. The somewhat different techniques evolved to deal with the two cases will now be treated separately.

4.2 In the first place consider the case of qualitative characters. There are a number of methods of illustrating a set of such data, four of which will be described here.

Example 4.1 This example makes use of the count of motor-vehicles given in table 3.1 (p. 24) and discussed in some detail in section 3.1. The data will be illustrated by four methods.

Method (i) A series of lines are drawn, the length of each line representing the number of individuals in one of the classes into which the data are divided. The base of each line is on the same level so that the heights reflect the differences between the numbers in the various classes. The result is shown in fig. 4.1 where each of

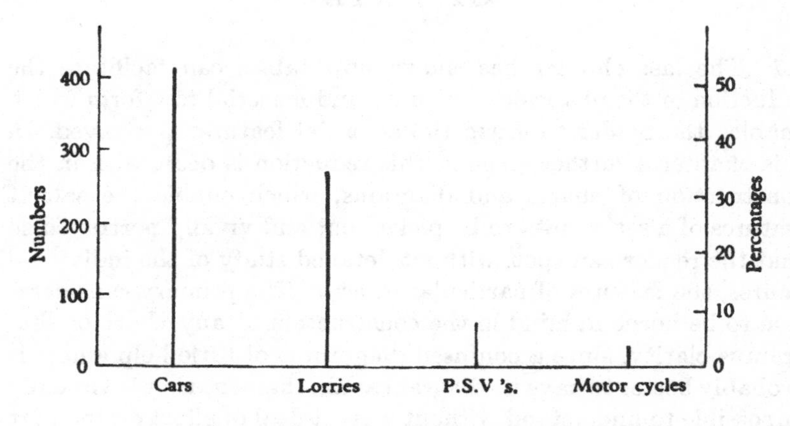

Fig. 4.1. Vehicles on the Great North Road

the series of lines is labelled with the type of vehicle to which it refers, and in height is proportional to the number of vehicles. To facilitate reading the figure and to increase its usefulness a scale is given on the left showing the numbers of vehicles represented by the vertical lines. Sometimes, as a further aid, an additional scale is given on the right, showing what percentage of the total number of observations each vertical line represents. Thus 52 per cent of the observed vehicles were cars. From such a diagram of this form, much information can be obtained without reference to the original data. The absolute numbers as well as the percentages are given in the diagram and there is no need to give a table as well as the diagram. It is, however, essential that the numbers and not just the percentages are given, so that information is not suppressed. It is customary to place the categories in descending order of magnitude as was done when such data were tabulated in tables 3.1 and 3.2. If there is a large number of categories and this rule is not followed, the resulting diagram can be extremely confusing to anyone looking at it for the first time.

Method (ii) As an alternative to drawing vertical lines a series of blocks can be drawn, each block being of equal width, but with heights proportional to the number of vehicles observed in each category (fig. 4.2). With the numerical and percentage scales attached this method is very similar to the previous one. Sometimes the blocks are shaded in order to make them stand out more clearly, and the block method is, therefore, often preferable to the line method, although the preference for one or the other is mainly a matter of personal choice.

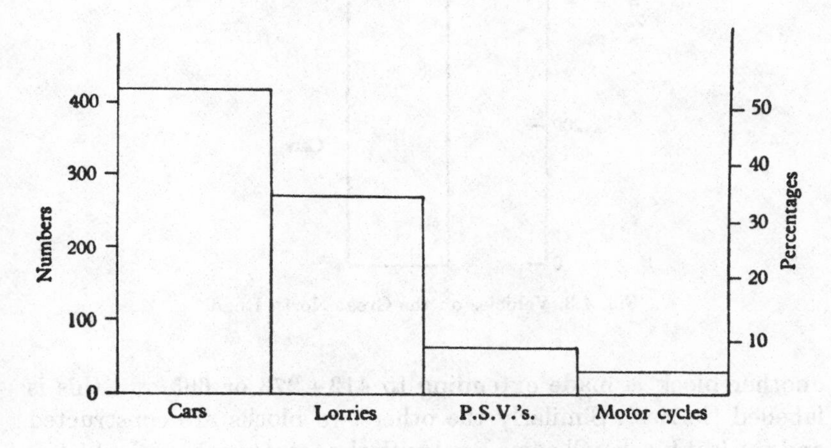

Fig. 4.2. Vehicles on the Great North Road

In both these methods it is possible to turn the diagrams through a right angle so that the base line is vertical and the lines or blocks representing the characters stick out horizontally. This system is more common with the line method than with the block method but the principles involved are exactly the same as before.

Method (iii) An alternative to the second method, the block method, would be to put the blocks on top of one another instead of side by side as previously. This forms what is called a *bar diagram* and is shown in fig. 4.3. It is constructed by drawing a line representing numbers and going from 0 to the total number of vehicles, namely 788. A block is then drawn on a convenient base level with the bottom of the scale and extending up to 412. This block is labelled 'cars'. Using the top of this block as the base

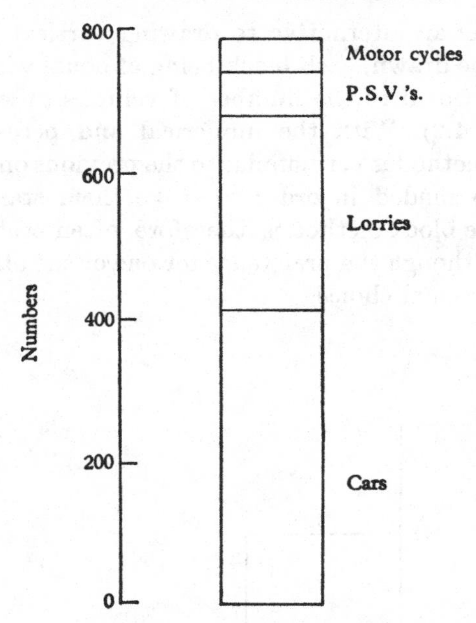

Fig. 4.3. Vehicles on the Great North Road

another block is made extending to 412 + 273 or 685 and this is labelled 'lorries'. Similarly the other two blocks are constructed and the last block will come level with the total number of vehicles, 788, on the scale. This is a very convenient form of diagram as it does not occupy very much space and a number of such diagrams can be placed side by side thus enabling very straightforward comparisons to be made between different distributions.

A common example of a bar diagram, which the reader is left to obtain, is that issued annually by the Treasury showing how the revenue that is obtained by the Exchequer each year is divided up into revenue from various sources such as income-tax, surtax, death duties, duty on tobacco and alcohol, purchase-tax and so on. At the same time another bar diagram is issued which shows the nation's expenditure divided into the expenditure under different headings such as defence, social services, pensions, health service, justice, housing and so forth.

Method (iv) The fourth method is to divide a circle of convenient size into segments, the area of each segment being proportional to the numbers in that category. Thus as there are 360 degrees in a circle the segment representing cars should subtend

$$\frac{412}{788} \times 360 = 188 \text{ degrees}$$

at the centre of the circle. Similarly lorries subtend 125 degrees and the other two categories take up the remaining 47 degrees. The resulting diagram is shown in fig. 4.4 and is quite a frequent form of straightforward diagram. Sometimes this, a circular form of chart, is referred to as a *pie chart*. It is difficult to show the actual numbers that have been observed in each category without a separate table and this is sometimes a drawback. This may be overcome by writing the numbers in the various segments, but it may make some of the smaller segments rather too full of writing to be read at all easily. It must be emphasized that in every case it is important to label the diagram so that it is easily understood. The reader can then get a firm grasp of the details without having to refer back to the original data, which may not be available.

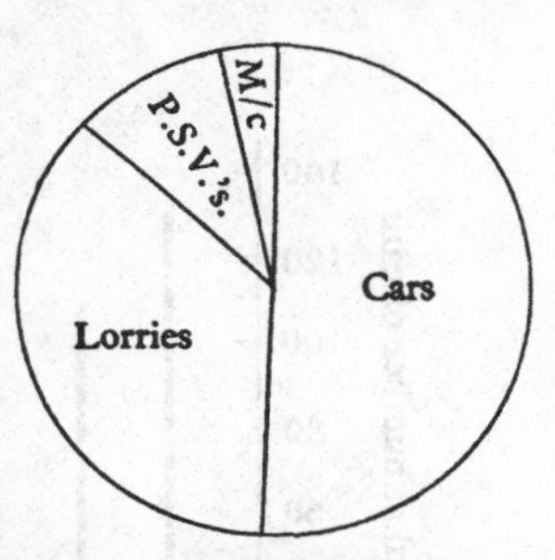

Fig. 4.4. Vehicles on the Great North Road

The choice of the most suitable method to use for any particular set of data depends on various considerations, such as the number of categories into which the data are divided, the space available for the diagrams, whether a series of comparative diagrams is required, and the worker's personal preferences. The overall object is to combine clarity with accuracy, and it should be noted that methods (i) or (ii) would show *small* differences between sets of data more clearly than (iii) or (iv).

4.3 The display of quantitative measurements must now be considered and here again a number of methods are available depending to some extent on the form of the data. First consider the case where the variable can only take certain distinct values.

Example 4.2 The data are the calls to a fire station over a year, given in table 3.4 (p. 26) and studied in some detail in section 3.3. For this data the variable, the number of calls to a fire station in a day, may take only positive integral values 0, 1, 2, ... The simplest method is to draw a line diagram analogous to that of fig. 4.1. In this case the categories correspond not to qualities but to the quantities 0, 1, 2, ... A base line is drawn on which 0, 1, 2, ..., etc., are marked out at equi-distant points. At each point a vertical line is drawn whose height is proportional to the number of days on which that number of calls were received at the fire station. The resulting diagram is shown in fig. 4.5 and it is easy to pick out the

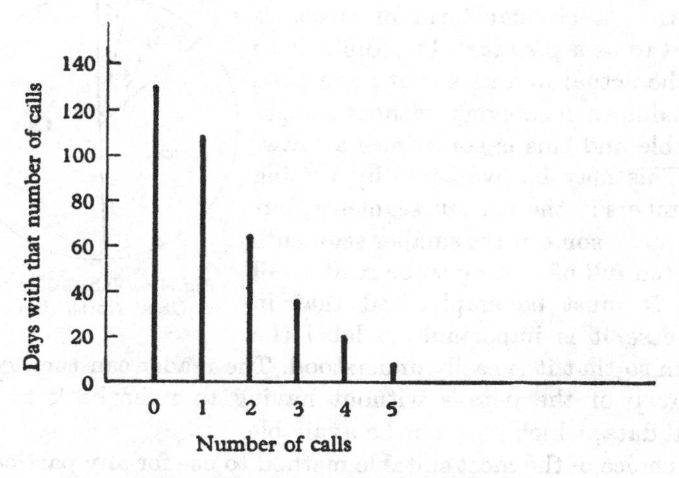

Fig. 4.5. Calls to fire station each day

relative proportions of days with few calls and those with many calls. Of course the order of the lines is fixed in this case and must go 0, 1, 2, ... the first line representing no calls, the second line just one call, the third line just two calls and so on. It would be misleading to place them in any other order. It so happens that in this example the method places the categories in decreasing order of magnitude, but it will not always be so and sometimes there will be tall lines followed by short lines followed by tall lines. This is preferable to altering the order of categories to say, 2, 0, 1 which would merely result in a very confusing form of diagram.

4.4 Next, consider the variable that is not only quantatitive but is also measurable on a continuous scale.

Example 4.3 Suppose that there are only a few observations available and take as an example the heights of ten schoolboys given in the first column of table 3.7 (p. 29). As there are only ten heights it is clear that only a simple method of representing these diagrammatically is possible. The first step is to draw a horizontal line to represent height. The lowest of the ten heights given is 59·2 in. and the largest 66·7 in. Hence the line must have a convenient scale that goes from about 59 to 67 in. When the line has been drawn and the scale inserted the first height is represented by a dot placed above the line opposite 63·3 on the scale. The second height is 60·0 and a dot is placed above the line opposite 60·0 on the scale. This process is now repeated for the other eight heights. The result is shown in fig. 4.6. It should be noted that had any

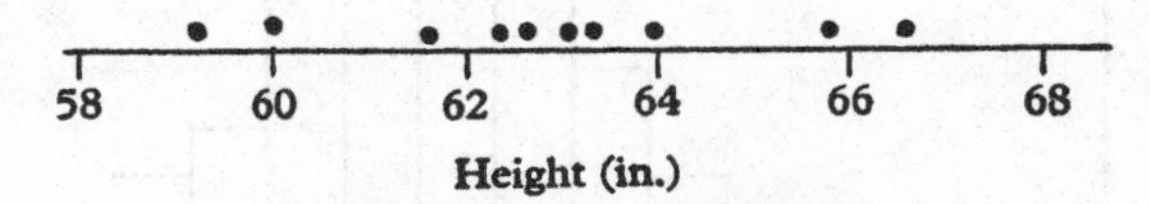

Fig. 4.6. Dot diagram of heights

of the heights been the same then the second dot would have been placed directly above the first one. The diagram is called a *dot diagram* and is quick and straightforward both to construct and to understand when the number of observations is small. An advantage of this method is that a number of small distributions can easily be compared by placing them one under each other in the form of a series of dot diagrams. Any big differences between the distributions are immediately obvious and no further analysis is needed. However, this method loses its usefulness when the number of observations is at all large, as not only is the diagram tedious to draw but the multiplicity of dots makes it far from pleasing to the eye and difficult to interpret. In such cases some other form of diagram must be used.

4.5 By far the best form of diagram is an extension of the blocks method (fig. 4.2) for the data concerning vehicles. In this example the data used will be more numerous than for the dot diagram.

Example 4.4 The data consist of the 100 schoolboy heights given in table 3.7 (p. 29). The base of each block corresponds to an interval on the measured scale of height. The data were given in the form of a table in table 3.9 (p. 29). A scale is now drawn reading from 54 to 69 in. and each inch marked off along the base. Using the appropriate portion as base, a rectangle is drawn whose height represents the number of boys in the second column of the table. The number scale is again put on the left-hand side and the top of the block is drawn level with the number of boys whose heights lie within the limits of the base of the block. The complete data of table 3.9 are shown in this manner in fig. 4.7. The small square

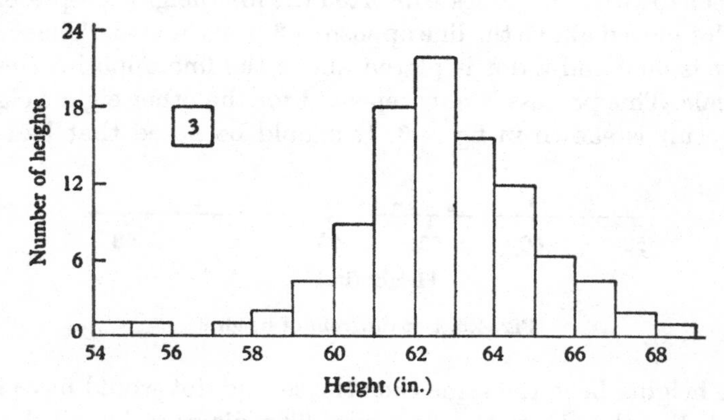

Fig. 4.7. Heights of schoolboys

with 3 inside it shows that this area represents three boys. Three times the number of these small squares in each rectangle in fig. 4.7 is the number of boys in that group. There is no need, of course, for the small square to represent three boys; three was chosen as being a reasonable size for the purpose of this figure, but a square one third the area might have been drawn to represent one boy. This form of block diagram is extremely common and very useful. It is simple to draw, neat, and easy to understand.

A slight modification of the block diagram sometimes met with is drawn by placing a dot at the point where the centre of the top of each rectangle would be. These dots are then joined up to produce a continuous line. Such a diagram is called a ***frequency polygon*** and conveys a very similar impression to the block diagram.

4.6 Difficulties sometimes arise if the group intervals used in the tabulation of the data are not equal.

Table 4.1. *Incomes of £2,000–£10,000 in Great Britain in the year ending 5 April 1965*

Income in £	No. of incomes
2,000–	95,000
2,500–	62,700
3,000–	67,900
4,000–	35,100
5,000–	21,701
6,000–	34,899
(but not over 9,999)	

Example 4.5 The data in table 4.1 are taken from an official publication, and give the numbers of persons with incomes between the limits of £2,000 and £10,000 divided into six intervals. It will be seen, however, that the six intervals or groups do not all have the same width but vary in width from £500 to £4,000. In fact three different intervals are utilised, namely £500, £1,000 and

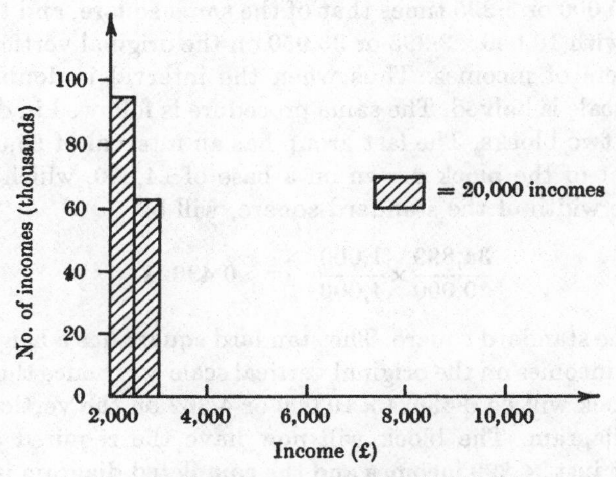

Fig. 4.8. Incomes in range £2,000–£10,000

£4,000. Some device must be used to overcome this difficulty, as the previous technique of having a scale on the left-hand side for use with all the blocks will produce rather different results according to the method of grouping that is adopted. The system used is to imagine some constant interval to be used throughout, and then to turn all the intervals into units of this constant interval. Two methods could be used to carry out this standardisation, both leading to the same diagram.

Method (i) Imagine that £500 is selected as the constant interval to be used throughout. It is in fact the true interval of the first two groups, and using it the horizontal scale is drawn from £2,000 to £10,000 and the vertical scale is drawn from zero to about 100,000. The first two blocks can now be inserted as before. Thus the first block has its base from £2,000 to £2,500 and its height is equal to 95,000 on the vertical scale. The resulting blocks are shown shaded in fig. 4.8. The next step is to draw a small square and to determine how many incomes it actually represents. In this case it will be seen that the area of the small square shown represents 20,000 incomes. The procedure now is as follows. Using £3,000–£4,000 as base draw a rectangle such that its area is 67,900/20,000 times the area of the small square, so that the area of the block drawn represents 67,900 incomes. Since the base of £3,000–£4,000 is the same as that of the small square the height has to be 67,900/20,000 or 3·395 times that of the small square, and this will be level with 10,000 × 3·395 or 33,950 on the original vertical scale of numbers of incomes. Thus when the interval is doubled the vertical scale is halved. The same procedure is followed in drawing the next two blocks. The last group has an interval of £4,000 and the height of the block drawn on a base of £4,000, which is four times the width of the standard square, will be

$$\frac{34{,}899}{20{,}000} \times \frac{1{,}000}{4{,}000} \quad \text{or} \quad 0{\cdot}43624$$

that of the standard square. The standard square has a height that is 10,000 incomes on the original vertical scale and hence the height of the block will be 0·43624 × 10,000 or 4,362 on the vertical scale on the diagram. The block will now have the required area to represent just 34,899 incomes and the completed diagram is shown in fig. 4.9.

Method (ii) An alternative and simpler method of constructing the blocks is to imagine that within each group the incomes are evenly spaced. Thus in the group £3,000–£4,000 there are 67,900 incomes and the basic unit of interval is £500. Hence divide the base into two portions £3,000–£3,500 and £3,500–£4,000 each of

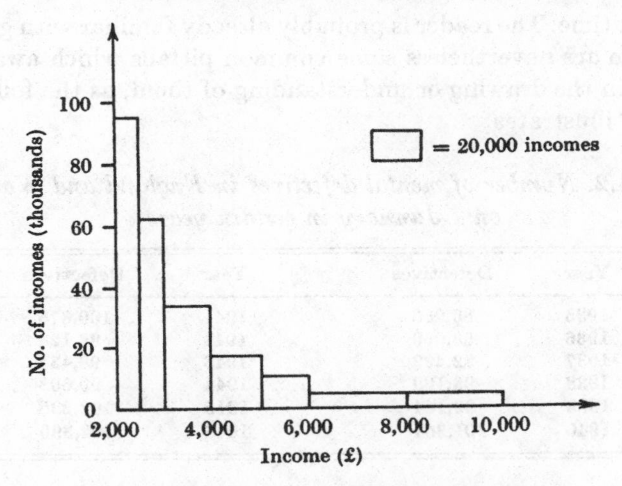

Fig. 4.9. Incomes in range £2,000–£10,000

width £500. Now if the 67,900 incomes are equally distributed over the whole range of £3,000–£4,000 then there will be 33,950 between £3,000 and £3,500 and another 33,950 between £3,500 and £4,000. If the two blocks corresponding to 33,950 are drawn it will be found that they are the same rectangles as before. Similarly the last group goes from £6,000 to £10,000 and could be divided into eight groups each of width £500 and with $\frac{1}{8} \times 34,899$ or 4,362 incomes contained in each of them. Thus the eight groups would be

£6,000–£6,500	4,362 incomes
£6,500–£7,000	4,362 incomes
£7,000–£7,500	4,362 incomes
⋮	⋮
£9,500–£10,000	4,362 incomes

If these eight blocks are now drawn using the original horizontal and vertical scales it will be found that the completed block diagram is exactly the same as before and hence the two methods are basically equivalent. The first method does in fact assume that the incomes are evenly spread over each group. This particular example emphasises the care which must be bestowed upon the drawing and interpretation of diagrams.

4.7 The question of scales becomes even more important when drawing graphs of some continuously measured variable over a period of time. The reader is probably already familiar with graphs, but there are nevertheless some common pitfalls which await the unwary in the drawing or understanding of them, as the following example illustrates.

Table 4.2. *Number of mental defectives in England and Wales on 1 January in certain years*

Year	Defectives	Year	Defectives
1935	86,086	1941	100,876
1936	88,060	1942	98,125
1937	92,299	1943	98,434
1938	96,109	1944	99,608
1939	99,144	1945	102,225
1940	101,364	1946	102,390

Example 4.6 Table 4.2 shows the number of registered mental defectives in England and Wales over the eleven-year period from 1935 to 1946. The information is shown graphically by two graphs in fig. 4.10. The impression obtained from the first graph is that there was a staggering rise in the number of mental defectives in little more than a decade, whereas the second graph gives the impression of a very much slower and more gradual increase. These

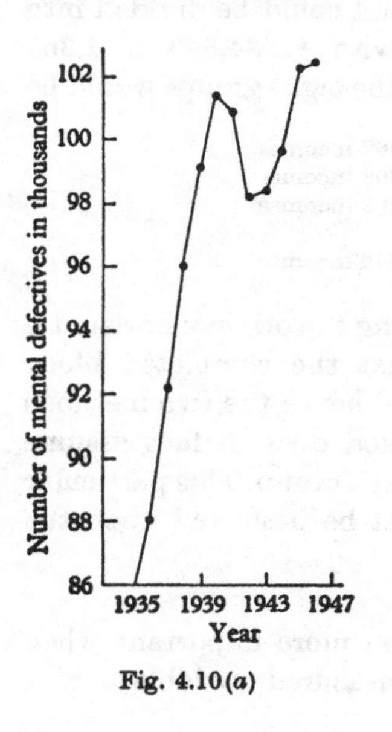

Fig. 4.10(*a*)

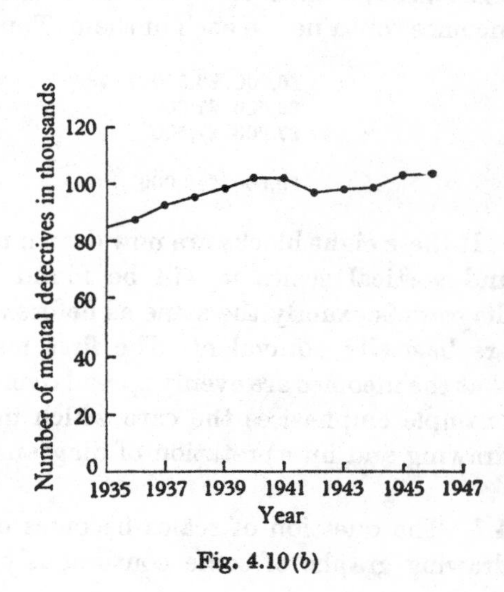

Fig. 4.10(*b*)

rather different impressions are obtained, of course, by tampering with the horizontal and vertical scales used. The vertical scale is large in the first case and the horizontal scale is large in the second case. It should be noted, however, that both the graphs are correctly plotted and both indicate the true numerical scales used. This shows that the scales of any graph must be examined before any conclusions are drawn from it; general impressions are not enough. If one or both scales do not start at zero but at some point above it (as in the vertical scale for fig. 4.10 (*a*)) then this fact must be clearly marked lest a misleading impression is given. It is often best to adopt a method such as that used in fig. 4.11, where the

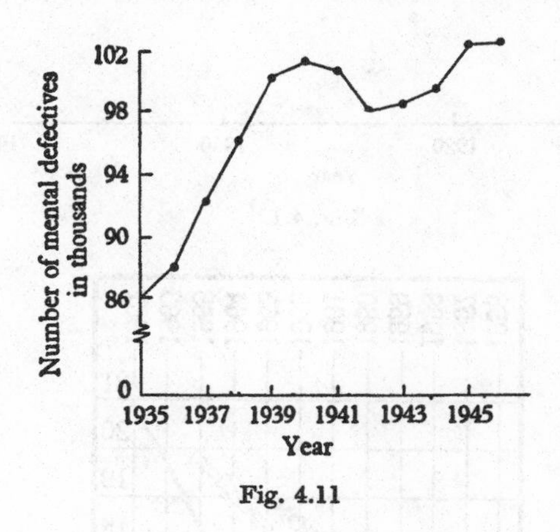

Fig. 4.11

zero is clearly indicated. Further it is essential to recognise the importance of supplying graphs, as well as tables, with full and clear labelling and if possible, the source of the information.

As another example of a misleading graph look at fig. 4.12 which purports to show the deaths per 100,000 population due to tuberculosis over some thirty years. The impression given is that of a slow decline over the years concerned. As the scale on the vertical axis, showing that in under 40 years the rate has been halved, from 100 down to 50, has been omitted, the reader is given only a vague idea of the progress of the decline. However, it is better than graphs with no units or scales on either of the axes!

4.8 Figure 4.13 is extracted from an article in a daily newspaper and may be taken as a fairly typical chart that appears in such

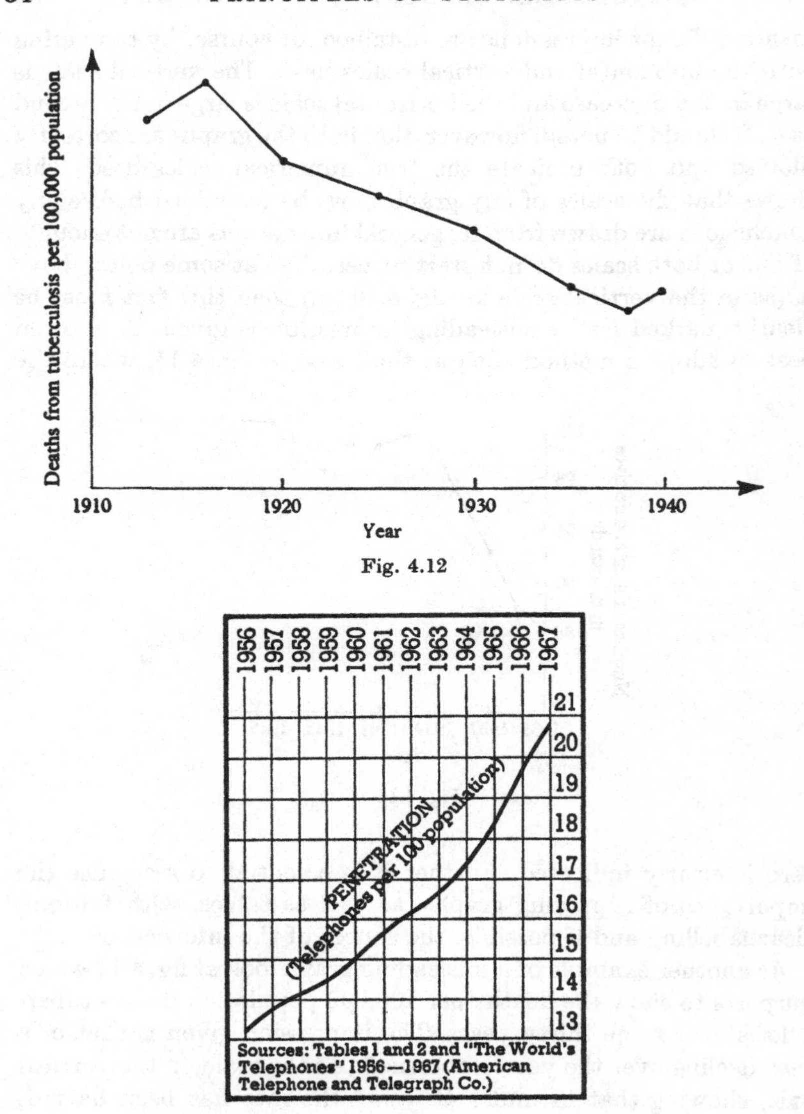

Fig. 4.12

Fig. 4.13. Telephone penetration in the U.K. 1956–67

papers. Notice that the chart is well defined; penetration is stated clearly and unambiguously. There are probably only two points that need comment. First, the scale is such that the rise over time looks more dramatic than it really is. The compounded rate of growth averages approximately $4\frac{1}{4}$ per cent per annum. Most readers would guess it to be somewhat higher. Secondly, the scale on the right hand side of the chart does not show precisely where 13 or 14 or 15, etc., actually fall. Is it on the line below, or mid-way between two consecutive lines?

4.9 In some problems, as has already been noted, two measurements are made on each individual. For instance, each of twenty schoolboys has his height and weight recorded. To represent such

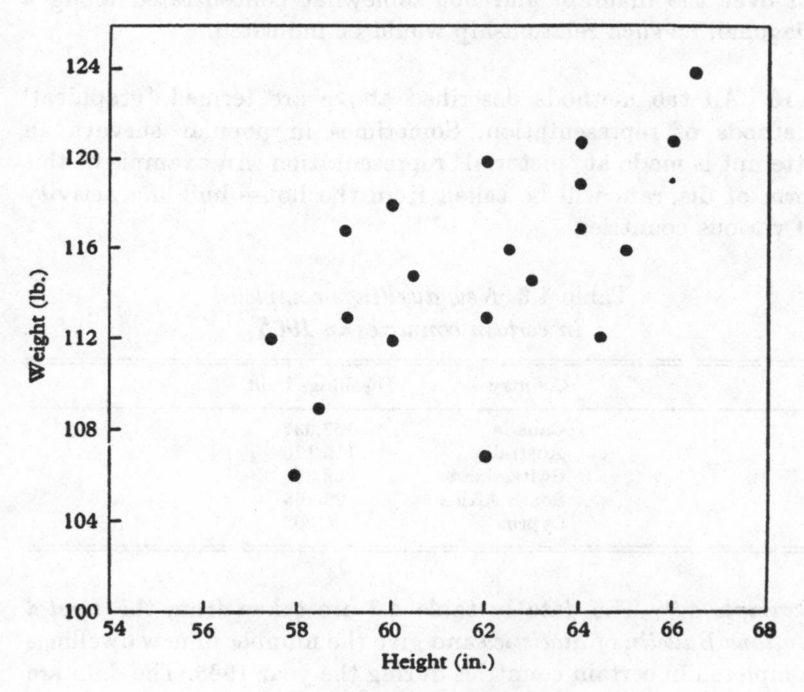

Fig. 4.14. Scatter diagram

data a two-way dot diagram can be formed. The horizontal scale is made to represent height in inches, and the vertical scale to represent weight in pounds. For each boy a dot is marked opposite the point on the horizontal scale representing his height and the point on the vertical scale representing his weight. For example, the dot in the bottom left-hand corner of fig. 4.14 represents a boy whose height is 58 in. and whose weight is 106 lb. Similarly the other nineteen dots in fig. 4.14 are placed to correspond with the boys' measurements.

This form of diagram is extremely valuable and is often referred to as a *scatter diagram*. It brings out clearly the relationship between height and weight, that is the heavier boys tend to be tall and the lighter boys short. This phenomenon is indicated by the manner in which the dots are approximately clustered along a diagonal running from the bottom left to the top right of the figure. If the dots had been scattered more or less haphazardly all over the diagram and not somewhat concentrated along a diagonal, no such relationship would be indicated.

4.10 All the methods described above are termed 'graphical' methods of representation. Sometimes in popular surveys an attempt is made at 'pictorial' representation. An example of this form of diagram will be taken from the house-building activity of various countries.

Table 4.3. *New dwellings completed in certain countries in 1965*

Country	Dwellings built
Canada	153,037
Australia	115,170
Switzerland	58,300
South Africa	22,464
Cyprus	4,092

Example 4.7 The data in table 4.3 are taken from the *United Nations Bulletin of Statistics* and give the number of new dwellings completed in certain countries during the year 1965. The data are shown in a pictorial manner in fig. 4.15 where one little house represents 20,000 dwellings built. This method of using a group of objects each of equal size, and comparing the size of the group rather than the absolute size of a block is sometimes a very forceful way of comparison. It is particularly common in advertising where a general rather than a detailed impression is often desired. For more accurate work however, such a method lacks precision, and hence the methods described earlier in this chapter are more frequently employed by the practising statistician.

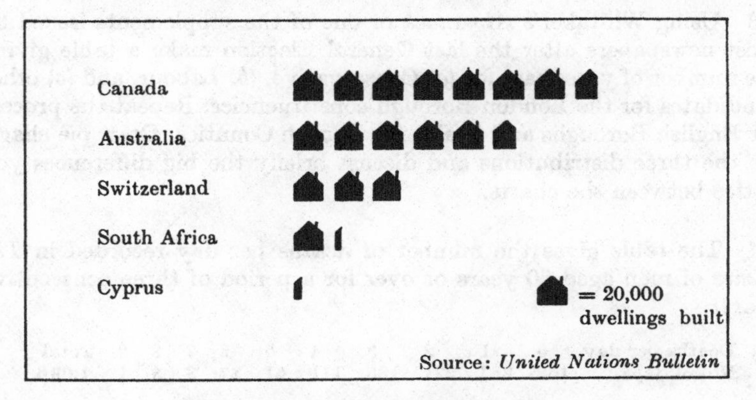

Fig. 4.15. Dwellings built in 1965

EXERCISES

4.1 The table gives the distribution of the working population of Great Britain in June 1959 and June 1966.

Employment	June 1959 (thousands)	June 1966 (thousands)
Armed Forces	565	417
Civil employment	23,242	24,974
Employers and self-employed	1,677	1,673
Unemployed	389	253
Totals	26,373	27,297

Represent this data diagrammatically in three distinct ways, and comment on the relative merits of the different representations.

4.2 The weekly receipts of the London Transport system for their three types of transport are given in the table for three periods.

	Railways	Buses and coaches	Trams and trolleybuses
April 1948	281	605	215
April 1950	275	589	204
April 1952	352	748	193

(All figures are in units of £1,000.)

Draw bar charts comparing the three years and comment on the big differences between 1950 and 1952.

4.3 Using Whitaker's *Almanack* or one of the supplements issued by most newspapers after the last General Election make a table giving the number of votes cast for (*a*) Conservative, (*b*) Labour, and (*c*) other candidates for the London Borough constituencies. Repeat the process for English Boroughs and finally for English Counties. Draw pie charts for the three distributions and discuss briefly the big differences you notice between the charts.

4.4 The table gives the number of deaths per day recorded in *The Times* of men aged 80 years or over for a period of three consecutive years.

Deaths per day	0	1	2	3	4	5	6	7	8	9	Total
No. of days	162	267	271	185	111	61	27	8	3	1	1,096

(*a*) Draw a diagram to represent this data.

(*b*) How many deaths of men of 80 years or over were recorded during the three years?

4.5 Two varieties of tomatoes are grown on twenty plots of equal size and the yields of tomatoes in kilo-grams are

Variety *A*	1·375	1·047	1·068	1·725	1·773
	1·201	0·779	1·042	1·223	1·633
Variety *B*	1·033	1·217	0·984	1·615	1·693
	0·673	0·840	0·842	1·252	1·217

(Data due to K. Mather.)

By plotting a dot diagram for each variety compare the yields obtained for the two varieties.

4.6 In the table below the numbers of children who passed G.C.E. O-level and A-level in each of eight consecutive years are given. Draw a graph to illustrate these figures.

Year	O-level passes	A-level passes
1958	748,184	122,315
1959	847,680	128,757
1960	928,028	146,362
1961	952,683	167,894
1962	1,050,851	182,236
1963	1,242,286	194,253
1964	1,264,259	215,598
1965	1,257,683	254,788

4.7 The table gives, in millions of pounds, the monthly figures for imports and exports from Great Britain. Represent these figures on a graph and comment on the result.

Date		Imports (£m)	Exports (£m)	Date		Imports (£m)	Exports (£m)
1967	Jan.	553	468	1967	Oct.	519	339
	Feb.	523	452		Nov.	574	340
	Mar.	499	432		Dec.	605	449
	Apr.	547	430	1968	Jan.	617	500
	May	578	437		Feb.	656	496
	June	511	417		Mar.	652	489
	July	493	432		Apr.	645	488
	Aug.	519	339		May	648	483
	Sept.	534	408		June	608	493

4.8 Draw a block diagram to represent the distribution of surtax payers in 1964–5 given in the table. How are you going to deal with the last group?

Income (£)	No. of incomes
2,000–2,500	95,000
2,500–3,000	62,700
3,000–4,000	67,900
4,000–5,000	35,100
5,000–6,000	21,701
6,000–8,000	23,299
8,000–10,000	11,600
10,000–12,000	6,701
12,000–15,000	5,600
15,000–20,000	4,030
Over 20,000	4,392

4.9 The table given below shows the Registrar General's estimate of the age distribution of the population on 30 June 1968. The figures are given in thousands.

Age last birthday	Population (thousands)	
	Male	Female
0–4	2,470	2,344
5–9	2,234	2,123
10–19	4,055	3,872
20–29	3,818	3,681
30–39	2,422	3,265
40–49	3,540	3,555
50–59	3,329	3,571
60–69	2,529	3,089
70–79	1,161	1,981
80 and over	351	806

(*a*) Draw a diagram to represent the figures.

(*b*) Comment on the difference between the male and female distributions.

(*c*) From the table it will be seen that there are considerably fewer males aged 30–49 than aged 10–29. What does this imply?

4.10 The table gives the number of deaths in 1947 of male and female children under 1 year old, subdivided according to the age at death.

Age	Deaths		Age	Deaths	
	Male	Female		Male	Female
Under 30 min.	430	434	1 week	1,440	1,148
30 min.–1 day	3,560	2,508	2 weeks	950	722
1 day	1,368	1,040	3 weeks	744	465
2 days	1,103	778	4 weeks–3 months	3,706	2,516
3 days	811	498	3–6 months	3,118	2,374
4 days	488	357	6–9 months	1,803	1,429
5 days	397	314	9–12 months	1,010	791
6 days	297	250			

The single age given is the lower boundary of the group.

Draw a block diagram for this data bearing in mind that there are six different intervals involved.

4.11 The degree of cloudiness may be measured on a scale from zero to ten. The table (due to G. E. Pearce) gives the degree of cloudiness at Greenwich during the month of July for the years 1890–1904 (excluding 1901).

Degree of cloudiness	No. of days	Degree of cloudiness	No. of days
10	676	4	45
9	148	3	68
8	90	2	74
7	65	1	129
6	55	0	320
5	45	Total	1,715

Draw a line diagram to illustrate these figures.

4.12 The figures for accidents involving motor cycles in column *A* below, are extracted from a report entitled *Road Accidents* published by H.M.S.O. The set of figures in column *B* represent the analysis of about 600 motor cycle accidents occurring abroad.

Object struck by motor cycle	Percentage of accidents	
	A	*B*
Car	25·1	39·8
Pedestrian	21·5	4·1
Property	18·9	7·2
Bicycle	14·7	4·4
Lorry	11·0	32·1
Bus	4·2	2·8
Motor cycle	3·1	7·5
Horse-drawn vehicle	1·5	2·1

Represent the figures in a diagram and write a brief report on any differences you can see between the two sets of figures. What rule has been broken in the method of presentation of the table?

4.13 **The table gives the employed percentage of trade union members and the marriage rate per 1,000 of the population (i.e. the number of marriages in England and Wales during the year expressed as so many per 1,000 of population) for thirty years. The data are taken from *Unemployment* by W. H. (later Lord) Beveridge.**

Year	Employed percentage	Marriage rate	Year	Employed percentage	Marriage rate
1900	97·55	16·0	1915	99·00	19·4
1901	96·65	15·9	1916	99·55	14·9
1902	95·80	15·9	1917	99·40	13·8
1903	95·00	15·6	1918	99·30	15·3
1904	93·60	15·2	1919	97·50	19·8
1905	94·75	15·3	1920	97·45	20·2
1906	96·30	15·6	1921	84·45	16·9
1907	96·05	15·8	1922	82·80	15·7
1908	91·35	15·1	1923	87·50	15·2
1909	91·30	14·7	1924	90·90	15·3
1910	94·90	15·0	1925	88·95	15·2
1911	96·95	15·2	1926	87·30	14·3
1912	96·85	15·6	1927	90·40	15·7
1913	97·90	15·7	1928	89·30	15·4
1914	96·75	15·9	1929	89·60	15·8

Draw a scatter diagram for employed percentage against marriage rate. From this diagram would you say that there is any truth in the assertion that a high employed percentage is associated with a high marriage rate and vice versa?

4.14 **In the table below are given the weights, in ounces, of the heart and kidneys of thirty men aged from 25 to 55 years. Plot a scatter diagram for the two weights. Is there any relationship between them?**

Adult no.	Weight		Adult no.	Weight		Adult no.	Weight	
	Heart	Kidneys		Heart	Kidneys		Heart	Kidneys
1	11·50	5·25	11	10·50	10·00	21	13·50	11·50
2	14·75	14·50	12	11·75	12·75	22	13·00	11·00
3	13·50	9·00	13	10·00	9·50	23	10·50	10·50
4	10·50	9·50	14	14·50	13·50	24	11·50	12·00
5	14·75	12·50	15	12·00	9·00	25	9·50	8·00
6	13·50	11·50	16	11·00	9·00	26	12·00	8·00
7	10·50	10·75	17	14·00	14·50	27	14·50	11·50
8	9·50	11·25	18	15·00	16·50	28	12·25	9·75
9	11·50	9·50	19	11·50	11·25	29	11·00	8·00
10	12·00	11·50	20	10·25	8·00	30	12·00	9·00

4.15 Measurements of span and length of forearm, in inches, were made on sixty adult men.

Adult no.	Span	Forearm	Adult no.	Span	Forearm	Adult no.	Span	Forearm
1	68·2	17·3	21	68·7	18·2	41	70·3	19·0
2	67·0	18·4	22	68·5	18·9	42	72·4	20·5
3	73·1	20·9	23	72·5	19·4	43	73·9	20·4
4	70·3	17·1	24	67·5	18·9	44	72·3	19·7
5	70·9	18·7	25	72·1	19·9	45	67·6	18·6
6	76·3	20·5	26	71·6	20·8	46	70·2	19·9
7	65·5	17·9	27	65·6	17·3	47	66·6	18·3
8	72·4	20·4	28	65·7	18·5	48	75·1	19·8
9	65·8	18·3	29	64·2	18·3	49	72·2	19·8
10	70·7	20·5	30	71·6	19·4	50	65·6	19·0
11	65·1	19·0	31	73·4	19·0	51	72·2	20·4
12	66·5	17·5	32	70·8	20·0	52	67·0	17·3
13	67·5	18·1	33	71·5	20·5	53	67·1	16·1
14	64·4	17·1	34	76·0	19·7	54	70·8	19·2
15	64·8	18·8	35	68·0	18·5	55	70·7	19·6
16	72·7	20·0	36	65·1	17·7	56	68·2	18·2
17	71·9	19·1	37	70·1	19·4	57	69·5	19·3
18	73·7	19·1	38	68·4	18·3	58	70·0	19·1
19	68·3	18·0	39	71·3	19·6	59	73·0	21·0
20	66·1	18·3	40	73·9	20·8	60	65·0	18·6

Form tables for span and forearm paying particular attention to the treatment of any observations which fall on the boundaries of your groups. Use a scatter diagram to discover whether the two variables, span and forearm, are linked.

4.16 The table gives the tensile strength of seventy-five malleable iron castings in pounds per square inch.

Tensile strength (central values)	Frequency	Tensile strength (central values)	Frequency
47,750	1	52,250	4
48,250	—	52,750	8
48,750	2	53,250	12
49,250	1	53,750	11
49,750	4	54,250	7
50,250	2	54,750	6
50,750	2	55,250	5
51,250	2	55,750	2
51,750	6	Total	75

Draw a block diagram to represent the data.

4.17 The table gives the numbers of deaths of persons involved in accidents on railways in England and Wales in 1952, either as a passenger or as an employee, subdivided according to age.

Age	No. killed	Age	No. killed
0–	9	40–	26
5–	2	45–	46
10–	2	50–	58
15–	39	55–	42
20–	46	60–	38
25–	27	65–	11
30–	35	Over 70	39
35–	29	Total	449

(*a*) Draw a block diagram to represent the data.

(*b*) Divide the data up into eight groups only, namely 0–, 10– and so on and draw a fresh diagram. Does it show any marked differences from that drawn under (*a*)?

4.18 The table gives the number of deaths per 1,000 live births of infants aged between 4 weeks and 1 year in the years 1921 and 1950, divided according to the social status of the father (data due to J. N. Morris and J. A. Heady).

	Deaths per 1,000 births	
Father's status	1921	1950
I. Professional	15·0	4·9
II. Intermediate	27·1	5·9
III. Skilled workers	43·2	10·5
IV. Partly skilled workers	52·7	14·1
V. Unskilled	60·2	17·9

There were 848,000 live births in 1921 and 697,000 in 1950. Draw diagrams to show (*a*) the differences in death rates according to status of father, (*b*) the change in death rates during the period.

5

FREQUENCY DISTRIBUTIONS

5.1 The reader who has conscientiously worked through the preceding chapter and its examples will have come across many charts and diagrams of varying shapes and forms. In order to refer easily to the form taken by such charts and diagrams, two technical terms will now be introduced.

Frequency distribution. The first term, 'frequency distribution', was briefly introduced in chapter 3 and can be illustrated by stating that a table such as table 3.9, where the numbers of schoolboys with heights in different categories are recorded, gives the frequency distribution of the heights of the 100 schoolboys. Thus a frequency distribution merely gives the frequency with which individuals fall into a number of different categories. The interval chosen for the classification is referred to as the *group interval*, and the frequency in any particular group interval is the *group frequency*. The manner in which the group frequencies are distributed over the group intervals is referred to as the frequency distribution of the variable.

Table 4.1 is a frequency distribution of the number of incomes that lie between £2,000 and £10,000 in the year ending 5 April 1965. The group intervals in this case do not remain constant over the whole range of the variable (income) but, nevertheless, the resulting distribution is still a frequency distribution. The exercises at the end of chapter 4 give numerous frequency distributions of variables, such as deaths per day, the tensile strength of iron castings, the ages of the population, and the degree of cloudiness at Greenwich. Their characteristic shapes may vary widely but they all give an indication of the spread of the variable over the range of variation that is being considered. Thus the frequency distribution can be regarded as the next stage in the experimenter's work after he has obtained his raw material and reduced it to manageable proportions.

Histogram. In chapter 4 the procedure by which frequency distributions could be represented by block diagrams was described in detail. The block diagram that corresponds to some particular frequency distribution is called the 'histogram' of the data. Thus fig. 4.7 gives a block diagram for the data of schoolboy heights in table 3.9 and is the histogram of the data.

Note that in a histogram a definite area represents a fixed number of observations. This fact enables the group interval used in a histogram to vary from group to group and yet the resulting diagram to give a true representation of the data. It is desirable when constructing any histogram to draw a small specimen area and label it with the number of observations that it represents. Then subsequent blocks can be drawn to represent the correct number of observations compared with this specimen area. The area on any base then represents the number of observations falling within the limits of the base. A histogram is, essentially, a useful method of illustrating any particular frequency distribution.

These two terms, frequency distribution and histogram, are important concepts, and will often be used and referred to in the following pages; it is therefore important to grasp fully the meaning of both terms.

5.2 In fig. 4.6 a dot diagram was drawn for the heights of ten schoolboys. The process can be continued and further dots added. Thus when a further ten boys have been measured, the twenty heights in all give a dot diagram such as in fig. 5.1. A further ten

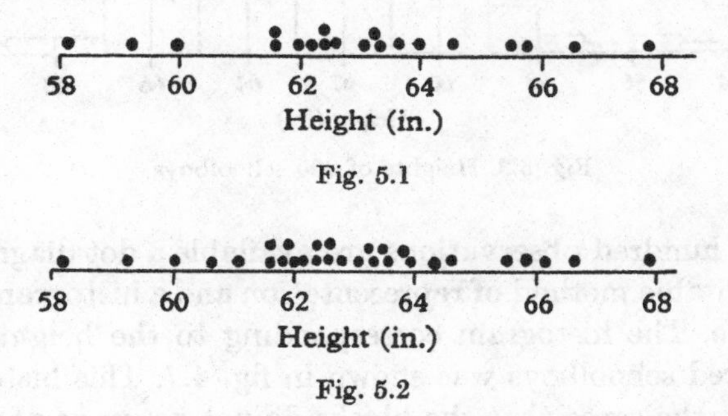

Fig. 5.1

Fig. 5.2

boys are now measured and the thirty heights now available give the dot diagram shown in fig. 5.2. This process of measuring more boys and putting in dots to represent their heights could clearly be continued indefinitely. The dot diagram would, of course, contain more and more dots as the number of schoolboys increased. It will be noticed that when there are ten dots only the pattern is very irregular and the dots do not appear to fall in any systematic form. A little basic information such as the limits between which the heights fall can be gleaned but not much else. When twenty schoolboys have been measured the twenty dots on the diagram are still in a very irregular form, although there is some small evidence of a pattern emerging, in that there are rather more dots in the centre of the spread than at either end. When the number of observations is increased to thirty the number of dots in the centre continues to grow at the expense of the two extremes, and the dots are spread in such a manner that, broadly speaking, a more definite pattern is emerging. Though it is still very crude and irregular, it leads us to expect that an even clearer pattern would emerge if the number of observations were still further increased.

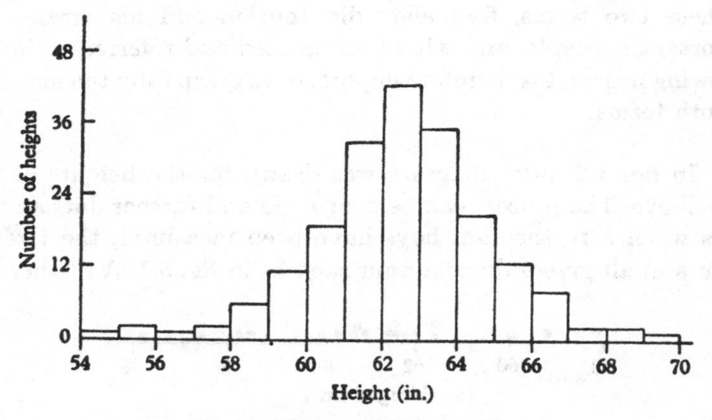

Fig. 5.3. Heights of 200 schoolboys

When a hundred observations are available a dot diagram is no longer a suitable method of representation and a histogram is more appropriate. The histogram corresponding to the heights of the first hundred schoolboys was shown in fig. 4.7. This histogram is irregular in the sense that the blocks do not go up or down in an exactly regular progression. The general impression gained from a study of the histogram is that the main bulk of the heights are clustered around 62 in., with a few spread out on either side, down to 54 in. at the lower end and up to 69 in. at the upper end. If 100 more schoolboys were selected and their heights measured a fresh histogram could be drawn, and the histogram resulting from the heights of the 200 schoolboys is shown in fig. 5.3. It will be noticed that the pattern is becoming more pronounced and regular but, owing to the fact that all heights within an inch are being grouped together, the histogram has a kind of 'step' appearance rising to a peak in the middle and then falling away again.

Suppose now that even more observations are taken and at the same time the observations are grouped together, not in groups that have a width or group interval of one inch, but in group intervals of half an inch. This will produce more 'steps', but each step will not be as high as before if the area under the histogram is kept about the same. This procedure has been followed using the heights of the same 200 schoolboys, the resulting histogram being based on a group interval of half an inch (fig. 5.4). It is rather

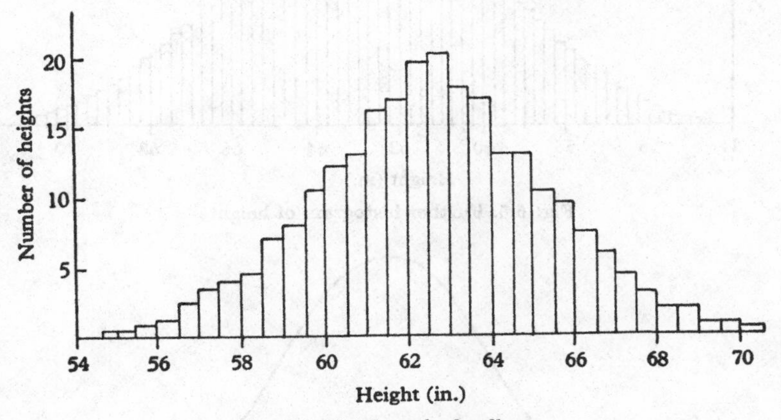

Fig. 5.4. Heights of schoolboys

smoother than before and gives a more regular appearance, with steps that are not too sudden or unusual. Each block shows a gradual change in height from the one that is next to it. This process of taking more and more observations and making the widths of the groups successively smaller could be continued. If necessary the heights would need to be measured to a greater degree of accuracy than that of one-tenth of an inch. A fresh histogram would approximate in shape to the previous one, but as the group intervals are decreased the steps become smaller and smaller until eventually the appearance of the histogram is closely akin to fig. 5.5. From here it is a simple matter to visualise that the limiting form of this histogram, as the group interval is decreased and the number of observations increased, will be the smooth curve drawn in fig. 5.6 which appears very little different from fig. 5.5, except that the vertical uprights of the blocks are now missing. This limiting curve is called a *frequency curve* and is

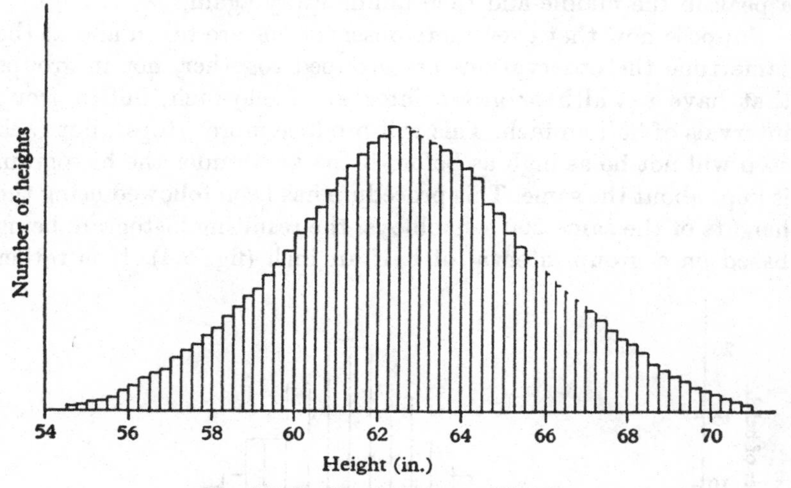

Fig. 5.5. Further histogram of heights

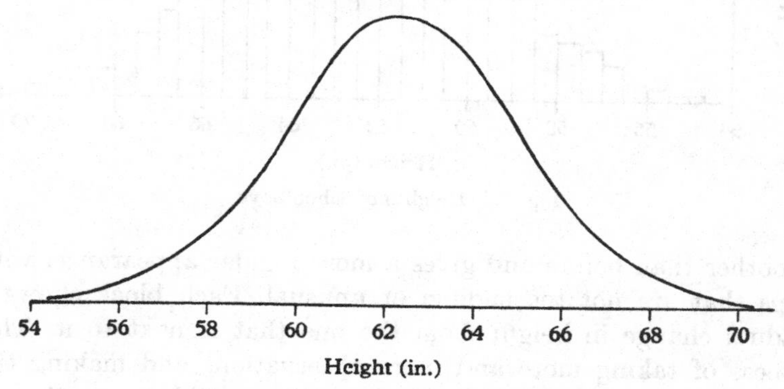

Fig. 5.6. Frequency curve of heights

of course analogous to the frequency distribution that was defined earlier, but for an indefinitely large number of observations. It must be strongly emphasised that this curve is the result of taking an extremely large number of observations and is an ideal that may be approached only if the number of observations is sufficiently large and their measurement sufficiently accurate.

5.3 The great value of a frequency curve is that it enables the properties of distributions to be examined, and varying forms of distributions to be described and compared in a general way. It is virtually impossible to produce a complete description of frequency distributions, as they can take so many different forms. But it is often possible to make an approximation to the form of the frequency curve underlying a particular distribution. By reference to this visualised frequency curve it is then practicable to make general comparisons between the distributions of two different variables or of the same variable in two different places. To illustrate how a frequency curve can vary it is proposed to discuss some of the forms taken by the frequency distributions of chapter 4. Five distinctive classes of frequency curves will be discussed in some detail. These curves by no means exhaust the possibilities, but they give a good indication of the types met with in practice.

Example 5.1 First there is the type of distribution which has been shown in fig. 5.6, representing the heights of schoolboys. It looks like a 'cocked hat' or a 'bell'. For brevity this type of frequency curve will be referred to in later pages as bell-shaped. It is very common in practice as many variables, such as height and weight of human beings and animals, and the measurements of plants, approximate to this form of distribution. A factory engaged in the manufacture of steel rods on a large scale, all of which are nominally of a certain fixed length, often finds in practice that the finished lengths vary slightly. No machine is perfect and the slight differences in the finished product will be due to the machine, the raw material, the temperature, the humidity, the machine-operator and so on. In such a case the final length of the rod is the nominal or planned length plus the effect of any errors due to the various factors involved. These errors may either be positive or negative and often they will cancel each other out. Sometimes, however, there will be more positive errors than negative ones, so that the final rod will be longer than the nominal length specified; or the reverse situation may occur. When a large number of rods have been manufactured and their lengths measured, the histogram of the lengths will usually be found to conform to a frequency curve which is of this bell-shaped variety. The nominal value of the length is in the centre of the bell, corresponding to the most common value, whilst the observed lengths of the manufactured rods are spread more or less evenly on either side of the central value, the curve decreasing to zero at either extremity.

This particular distribution or frequency curve is commonly referred to as a Gaussian or *normal* distribution and occupies a central position in the theory and application of statistics. Accordingly it will be discussed in more detail later, when various statistical tests are derived and illustrated. It must be borne in mind throughout, however, that any frequency curve is an ideal that can be reached only by a very large number of observations. Unless this is accepted a slight unevenness in the histogram could obscure the underlying frequency curve.

5.4 *Example* 5.2 The next form of frequency curve to be discussed has a very different shape. To understand how it can arise imagine that it is possible to measure the heights of schoolboys, all aged 14 years, very accurately indeed, say to five places of decimals (or a hundred thousandth of an inch). In using the measurements the figures are, as in the earlier example, rounded off to one place of decimals. By doing this a small error of recording is made. This is the difference between the accurate, or true height, and the 'rounded-off' height. This difference will lie between $-0{\cdot}05$ and $+0{\cdot}05$ of an inch and is often referred to as the rounding off error. If this error is obtained for a large number of boys—and it will require very accurate measurement on each boy—the tabulated errors will not have the same form as in example 5.1. The reason for this is that practical experience has shown that people's heights are not in general an exact number of inches. People with heights between, say, 62·15 and 62·25 in. are not in general concentrated at 62·2 in. exactly, but are more or less evenly spread over the range from 62·15 to 62·25 in. Inches are, of course, a purely arbitrary and man-made unit, and another unit such as centimetres could equally well be used. Even if boys always had exact heights on one scale of measurement they would have inexact heights on another scale. Thus it seems that the number of boys whose heights have the figure 3 in the second place of decimals will be about the same as the number having 7 in the second place of decimals, since there is no reason for one figure to appear more often than any other. This will be true for any pair of figures, so that an examination of a table of rounding-off errors would show them approximately equal spread between the limits of $-0{\cdot}05$

and +0·05. If more observations were taken and more rounding-off errors accumulated, the regularity of the corresponding histograms would become more pronounced and tables with equal group widths would have approximately the same number of observations in each group. Hence it seems that the frequency curve corresponding to such a table will be a level, or horizontal, line of uniform height between −0·05 and +0·05 in. Diagrammatically it is illustrated in fig. 5.7 and the distribution is often called the *rectangular distribution* as its graph resembles an ordinary rectangle.

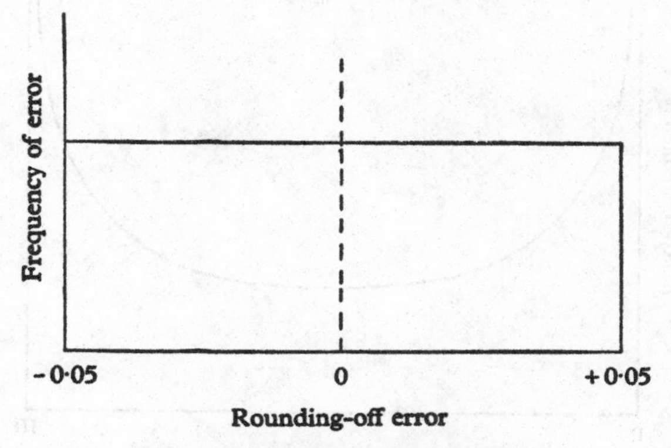

Fig. 5.7. Rectangular distribution

The rectangular differs from the bell-shaped frequency curve in two main features. First, the two ends of the distribution are firmly fixed at −0·05 and +0·05 in., whereas for the bell-shaped distribution there is no definite starting- or end-point, but a gradual tapering down to zero frequency at either end of the distribution. Secondly, instead of having a maximum, or most popular value, which occurs in the centre of the bell-shaped curve, the rectangular curve has all values equally likely and no single value occurs more often than any other.

5.5 In the bell-shaped distribution of example 5.1 the value at the centre was greater than the values on either side, whereas in the rectangular distribution there is no peak. The following example is of a distribution having a trough and not a peak in the centre.

***Example* 5.3** In exercise 4.11 of the last chapter a line diagram was drawn of the degree of cloudiness at Greenwich for the month of July over some fourteen years. The cloudiness was expressed on a scale which went from 0 up to 10. The line diagram gave the broad appearance of a large or capital U and, had the degree of cloudiness been expressed on a continuous scale and more observations become available, it is likely that a rather smoother curve would have been obtained. Thus the underlying frequency curve is probably something like that illustrated in fig. 5.8.

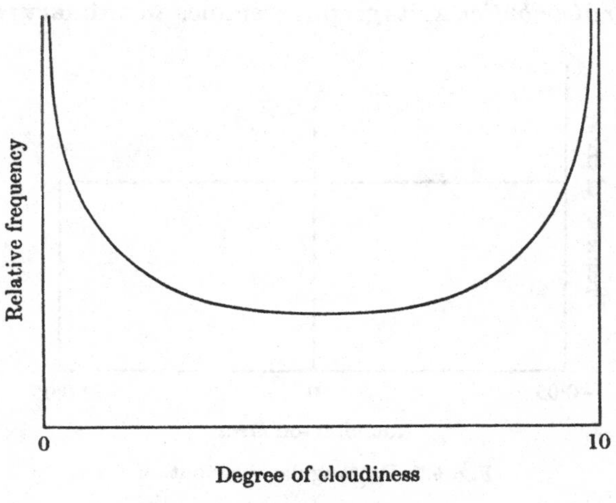

Fig. 5.8. U-shaped distribution

Expressed in words this means that there were many days when the sky was either completely clear or completely clouded but that few days were in between these two extremes. Once again the curve has its two ends or terminals fixed, but instead of having a maximum in the centre like the bell-shaped curve, it has a minimum value in the centre and a maximum value at either end. This form of frequency curve is in very marked contrast to the bell-shaped frequency curve. It rarely occurs in practice.

The three distributions discussed so far have been of a symmetrical form. By this it is meant that if a vertical line is drawn through the centre of the distribution the left-hand side is the mirror image of the right-hand side and vice versa. This is demonstrated in fig. 5.7 where the centre line is indicated as a dotted line. Any frequency that is a given distance to the left of the centre is matched by an equal frequency the same distance to the right.

5.6 Not all distributions are of symmetrical form and the next two examples are of non-symmetrical distributions or, as they are usually called, *asymmetrical* or *skew* distributions.

Table 5.1. *Ages of males married in England and Wales, 1965*

Age	No. married
16–20	65,146
21–24	175,281
25–29	94,468
30–34	30,554
35–44	26,910
45–54	13,518
55 and over	16,162
Total	422,139

Example 5.4 The basic data used for this example are the ages at marriage of males in England and Wales for the year 1965, given in tabular form in table 5.1. The data can be plotted as a histogram and it is at once apparent that the distribution is not symmetrical. The numbers seem to rise to a peak around 24 years of age, but there is a very rapid rise to this peak from below with a much slower dying away of the numbers when above 24 years. This is because, although marriages can legally take place at any age over 16 years, and some marriages take place right up to the age of 80, the most popular ages for marriage are from 22 to 27 years. Thus the lower terminal is some eight years below the peak but the upper terminal is very much farther away. If the age-groups were made narrower, the resulting frequency curve would be something like the very asymmetrical form of curve shown in fig. 5.9. In such a case there is still a maximum value, but it is no longer in the centre of the distribution. In this particular case the long tail occurs to the right of the maximum value, but this is not always necessarily so. For example, readings of barometric height taken daily over a long period at one place can give rise to a form of frequency distribution or curve that has the longer tail on the left-hand side.

5.7 The amount of skewness of the curve can vary a great deal. Consider a symmetrical distribution with the maximum value in the centre. To obtain a skew distribution the maximum is

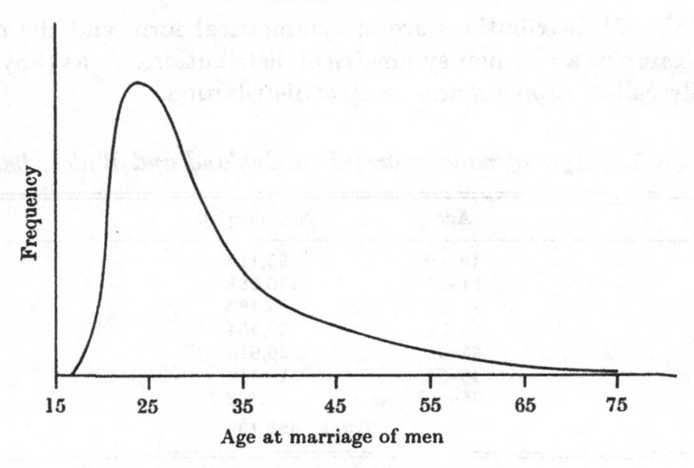

Fig. 5.9. Skew distribution

displaced from the centre to one side or the other. If the displacement is increased there comes a time when the maximum value is at one end and all the other values are less than the value at that terminal.

Example 5.5 Consider the data relating to incomes between £2,000 and £10,000 given in table 4.1 and illustrated with a histogram in fig. 4.9. Here every frequency is less than the frequency in the previous group, after allowing for differences in group intervals, and the first group has the largest frequency. The frequency curve that might be appropriate to such data is illustrated in fig. 5.10, and this form of curve is commonly referred to as a J-*shaped* curve. It is often met with in economic statistics, as it seems to be the frequency curve most appropriate to the distribution of incomes, rateable values of properties and so on. Extreme care is often necessary as there may be a maximum value which is not quite at the terminal but whose presence is masked by the method of grouping the data. For this reason it is a wise precaution, when selecting data that seems to come from a J-shaped distribution, to use a rather finer interval for the part near the maximum than for the other parts of the distribution.

5.8 The frequency curve is a useful concept, as it enables the broad outlines of a distribution to be discussed and comparisons

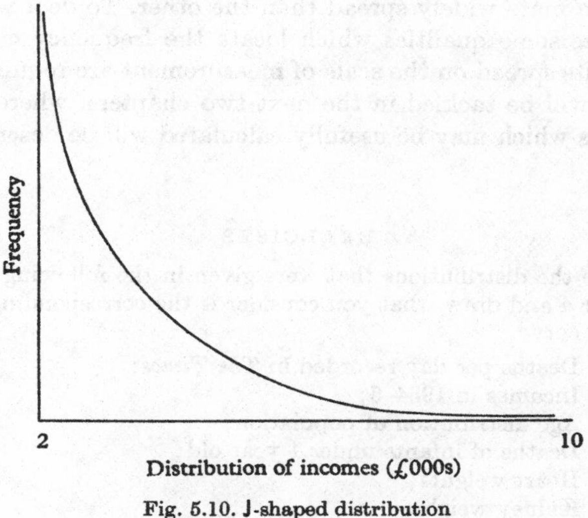

Fig. 5.10. J-shaped distribution

to be made without going into a lot of detail. The curves described here are only indicative of the broad range of possibilities. The three most common forms met with in practice are (1) the symmetrical or bell-shaped, (2) the moderately skew, and (3) the extremely skew or J-shaped, curves. The frequency distributions observed will not always fall exactly into one or other of these types, which are by no means exhaustive, and many practical examples will differ from the curves discussed here. In the following chapters a great deal of emphasis will be placed upon the special bell-shaped distribution mentioned in section 5.3. This arises quite naturally as the true frequency curve for many variables and is approximately true for others. It thus occupies a central position in statistical theory.

Although this concept of a hypothetical limiting frequency curve underlying any variable has enabled some rough and ready comparisons to be made between frequency distributions, there is still a need for further concepts that will summarise and represent the data with greater detail. For instance, the heights of two groups of schoolboys may give rise to frequency curves of exactly the same shape and yet all the boys of one group may be 3 in. taller than the corresponding boys in the other group. Or again, the general shape of two distributions may be the same but one may seem more widely spread than the other. To deal with such differences some qualities which locate the frequency curve and describe its spread on the scale of measurement are required. This problem will be tackled in the next two chapters, where various quantities which may be usefully calculated will be described.

EXERCISES

5.1 Take the distributions that were given in the following exercises of chapter 4 and draw what you consider is the corresponding form of frequency curve:

4.4 Deaths per day recorded in *The Times*;
4.8 Incomes in 1964–5;
4.9 Age distribution of population;
4.10 Deaths of infants under 1 year old;
4.14 Heart weights;
4.14 Kidney weights;
4.15 Span;
4.15 Forearm length.

5.2 Select a page of the telephone directory that is free from advertisements and count up how many telephone numbers have 0 as the last digit. Carry out the same procedure for telephone numbers ending in 1 and so on, up to those ending with 9 in the last digit. What form of frequency distribution would you expect to get, and does your actual distribution conform to this? If not, discuss reasons why it does not do so (cf. exercise 2.24).

5.3 Repeat the previous exercise this time taking the last digit of the registration number of cars passing along a road. See whether this distribution varies in shape from that of the last exercise.

5.4 In co-operation with your physics department arrange to keep a record of the rainfall per day over a period of about two months (60 days). Draw a histogram from the results and hence deduce the form of the underlying frequency curve.

If possible repeat the whole procedure at a different time of year and see whether the form of the frequency curve has changed at all.

5.5 Collect the scores obtained on the miniature range by members of the C.C.F. contingent at the school and form a frequency distribution, deducing from it the form of frequency curve that seems appropriate.

5.6 Examine the score-book of the school cricket XI and compile a frequency distribution of the number of runs scored in each over off some bowler. From this, attempt to find out the form of underlying frequency curve that seems to be the most realistic, remembering that you are here dealing with a discrete and not a continuous variable.

5.7 Lists giving the rateable values of houses are available in local Council offices. Using these lists compile a frequency distribution, for one ward of a borough, of the rateable values. A suitable grouping to take would probably be a group interval of £10. From this frequency distribution suggest the possible underlying frequency curve. Repeat the procedure for another ward in the borough (if possible one that has a rather different character from the first) and see whether the underlying frequency curve is different, and, if so, in what ways it differs.

5.8 Select a sample of 100 students or colleagues. Find out by questionnaire how long they took (*a*) to get to work and (*b*) to get home from work on the previous day. Plot both sets of data as histograms and attempt to sketch the underlying frequency curves. Do any differences emerge between the two sets of data?

6

AVERAGES

6.1 The preceding chapters have brought the investigation of a statistical problem to the stage where the data have been collected in an observer's notebook, reduced to more manageable proportions by means of tables, and the salient features portrayed by means of diagrams. The next stage is to attempt a summary of the data by the calculation of a few representative values that size it up and enable comparisons to be made swiftly and accurately between one set of data and another. Just as a town elects a Member of Parliament to the House of Commons at Westminster to represent its views and opinions, or a company may calculate its percentage return on its assets, so averages are selected in order to represent a given set of data.

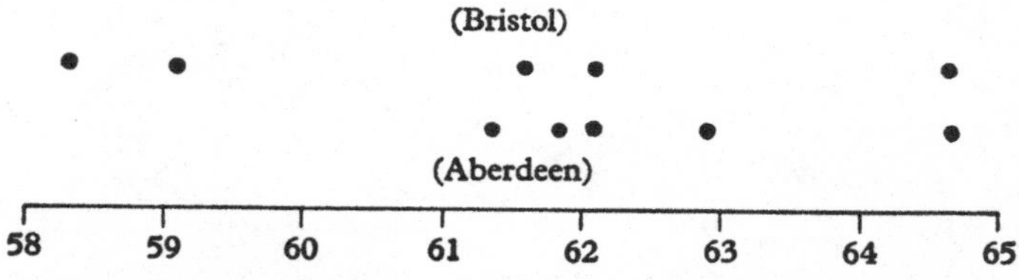

Fig. 6.1. Heights of schoolboys from Bristol and Aberdeen in inches

For example, imagine that five schoolboys, aged 14 years, have been chosen from amongst those in Bristol and that their heights, measured in inches, are

62·1 59·1 64·7 58·3 61·6.

A precisely similar procedure is carried out in Aberdeen and the five heights obtained are

61·4 62·1 61·8 64·7 62·9.

The question asked is whether there are any differences between the two series of heights. A dot diagram given in fig. 6.1 brings out the fact that the Aberdeen heights are slightly greater on the whole than the Bristol heights. What is now needed is some overall measure to describe the general level of the observations, and for this the most commonly employed quantity is the *mean* or arithmetic average of the observations.

The calculation of the mean is a straightforward operation. The individual observations are added up and the total thus obtained is divided by the number of observations. Each individual value plays an equal part in the determination of the mean. Carrying out this procedure gives the following results:

Bristol. Sum of observations:

$$62{\cdot}1+59{\cdot}1+64{\cdot}7+58{\cdot}3+61{\cdot}6 = 305{\cdot}8$$

Number of individuals: 5

Mean $= \frac{1}{5}(305{\cdot}8) = 61{\cdot}16$.

Aberdeen. Sum of observations:

$$61{\cdot}4+62{\cdot}1+61{\cdot}8+64{\cdot}7+62{\cdot}9 = 312{\cdot}9$$

Number of individuals: 5

Mean $= \frac{1}{5}(312{\cdot}9) = 62{\cdot}58$.

It will be noticed, as indeed was expected, that the Aberdeen boys have a higher mean height than the Bristol boys. The mean provides a measure of the general level of heights and acts as a kind of précis of the data. Naturally information is lost by giving only the mean of a set of numbers instead of the numbers themselves, but nevertheless it is often very valuable to give an idea of the general level of some variable, and the mean is readily available for this purpose.

In such comparisons it does not matter if the two groups to be compared have unequal numbers of observations, as the method of obtaining the mean eliminates such differences. Suppose in Aberdeen seven boys were measured and not five as before and the two additional heights were 63·6 and 61·7. The revised mean will now be $\frac{1}{7} \times 438{\cdot}2$ or 62·60 and this figure can still be legitimately compared with the figure of 61·16 for Bristol.

6.2 The mean is extremely simple to calculate, but in order to avoid large numbers and heavy work there are various short cuts that are often employed in the calculation. First, if all the values are large, a constant amount can be subtracted from each of them. The mean is then found from the modified values and the constant amount finally added on again. Thus if 55 were subtracted from the five Bristol heights of section 6.1 they become

7·1 4·1 9·7 3·3 6·6.

The mean of these five values is 6·16 so that the mean of the original values will be 55 + 6·16, or 61·16 as before. In this case there is very little saving but when there are many more observations the saving is much greater.

Example 6.1 Suppose that the mean height of the hundred schoolboys given in table 3.7 (p. 29) is required. A constant amount of 60 is first subtracted from each height giving table 6.1. It will be noticed that as some of the heights were below 60 in. they now appear as negative quantities. This does not in any way affect the argument and ensures that the numerical magnitude of the revised quantities is kept as low as possible.

Table 6.1. *Heights in arbitrary units*

3·3	−1·9	3·8	1·9	5·4	2·3	1·4	6·9	3·0	0·1
0·0	4·6	1·1	2·9	0·6	−0·7	5·2	1·2	−2·2	2·4
5·9	2·0	6·1	−1·1	4·1	4·2	1·3	0·8	7·0	5·0
3·1	3·7	1·8	4·3	2·6	−5·7	3·2	3·4	1·8	4·7
−0·8	2·3	0·7	5·6	1·1	3·8	0·5	2·8	4·1	1·9
4·0	1·7	4·5	1·3	0·2	1·2	6·3	−0·6	8·3	2·3
6·7	5·7	2·8	4·9	2·7	2·2	1·9	2·6	3·9	4·5
2·4	7·9	3·4	−4·3	3·1	−0·3	4·8	5·8	0·4	2·8
1·7	2·4	1·9	3·8	1·6	2·1	3·0	4·4	2·0	1·5
2·5	3·2	2·1	2·8	3·4	2·9	0·3	2·1	1·3	3·9

The sum of the positive numbers is 281·1, and that of the negative numbers − 17·6. Hence, in the arbitrary units which have been adopted, the mean is

$$\tfrac{1}{100}(281{\cdot}1-17{\cdot}6)$$

$$= \tfrac{1}{100}(263{\cdot}5) = 2{\cdot}635.$$

It is now necessary to allow for the fact that before calculating the mean, 60 was subtracted from each height. A moment's reflection shows that this will make the mean value in the arbitrary units 60 in. too small. Hence the true mean value is

$$60+2{\cdot}635 = 62{\cdot}635 \text{ in.}$$

This may be verified by adding together the original values in table 3.7. The general rule is that if a constant amount is subtracted from every observation to form a new series of observations, then the mean of the original observations is the mean of the new series plus the constant amount that has been subtracted from each observation. Clearly care should be taken in the selection of the constant amount that is subtracted. A glance at the data will usually give a rough idea of the magnitude of the mean, and if the constant amount subtracted is near this approximate mean, the magnitudes of the new values will be kept down to a minimum and hence simplify the arithmetic. It will, of course, make a significant proportion of the revised values negative, but this is no handicap.

6.3 The data whose mean is required will not, however, always be available as a series of individual values. Often the only information available will be in the form of a grouped frequency distribution. Thus instead of the hundred individual heights of table 3.7 the only information available might be the grouped frequency distribution given in table 3.10. From this table it is required to calculate the mean height.

It is clear that by putting the observations into groups a small amount of information has been lost. For instance, from the table twelve boys had heights between 63·95 and 64·95 in. but their exact heights are no longer known. Thus some assumption has to be made as to how the heights are spread within the group, and the most straightforward assumption is that the heights are spread evenly. If there are only a few groups, so that each group covers a very wide range of the variable, this assumption is not very good; but, provided that there is a reasonable number of groups, it should not lead to any appreciable error. If the observations are evenly spread over a group this is equivalent, for the purposes of calculating a mean, to their being concentrated at the central value of the group, since any values a given amount greater than the central value are exactly counterbalanced by an equal number of values the same amount below the central value. Hence the contribution to the mean of the combined set of values is identical with their contribution if they had all been at the central value. Making this assumption, the mean value deduced from table 3.10 will be

$$\tfrac{1}{100}(1 \times 54{\cdot}45 + 1 \times 55{\cdot}45 + 1 \times 57{\cdot}45 + 2 \times 58{\cdot}45 + \ldots$$
$$+ 2 \times 67{\cdot}45 + 1 \times 68{\cdot}45) = \tfrac{1}{100}(6262) = 62{\cdot}62 \text{ in.}$$

It will be noticed that in this calculation the labour is reduced by writing $2 \times 58{\cdot}45$ for $58{\cdot}45 + 58{\cdot}45$ and so on for each of the other groups. The value 62·62 obtained here should be compared with the value 62·635 obtained by using the original full set of values.

In general the larger the number of observations and the finer the interval of grouping, the more accurate will be the final mean calculated from the grouped distribution.

Table 6.2. *Calculation of mean*

(1) Height (in.) central values	(2) Arbitrary units	(3) No. of schoolboys	(4) Product (2) × (3)	
54·45	−7	1	−7	
55·45	−6	1	−6	
56·45	−5	0	0	
57·45	−4	1	−4	−40
58·45	−3	2	−6	
59·45	−2	4	−8	
60·45	−1	9	−9	
61·45	0	18	0	
62·45	1	22	22	
63·45	2	16	32	
64·45	3	12	36	
65·45	4	7	28	157
66·45	5	4	20	
67·45	6	2	12	
68·45	7	1	7	
	Total	100		
			Final total	117

6.4 The arithmetic of the calculation for a grouped distribution can, however, be simplified further by making use of what is termed an arbitrary origin, which is equivalent to the previous method of subtracting a constant amount. To use this principle a group in the centre of the distribution is chosen and nominated as zero. In this particular instance the group with central value 61·45 in. has been taken. The groups on either side are then re-labelled using the group 61·45 in. as a new origin of measurement. Thus the next group above, 62·45 becomes +1, the group above that, 63·45, becomes +2 and so on. Similarly below the centre the group 60·45 becomes −1 whilst the group 59·45 becomes −2 and so on. These are used as arbitrary units and are shown in column (2) of table 6.2. Column (3) gives the number of boys within each group and is taken direct from the original table.

The next step is to calculate the mean of the distribution in the arbitrary units. Using the method adopted earlier this would be equal to

$$\tfrac{1}{100}(1\times(-7)+1\times(-6)+1\times(-4)+\ldots+2\times 6+1\times 7)$$

and will thus require the sum of the products of column (3) with the corresponding item in column (2). Accordingly a fresh column (4) is formed giving these products. Column (4) is best summed in two parts, the negative part followed by the positive part. These two results are added together giving, in this case, 117. Hence the mean in the arbitrary units is $\tfrac{1}{100}(117)$ or 1·17. Now zero in the arbitrary units corresponds to the value of 61·45 in the original units. Hence the mean in the original units will be

$$61{\cdot}45 + 1{\cdot}17 \text{ or } 62{\cdot}62 \text{ in.},$$

which agrees with the result already obtained.

6.5 The larger the number of observations involved the more powerful becomes the use of an arbitrary origin. If there are not equal group intervals throughout the distribution, the arbitrary units will falsely represent the groups. In the example just discussed the group interval was one inch throughout. In cases where the group interval is not one unit the method has to be slightly modified, as shown in the next short example.

Example 6.2 The data in table 6.3 give the yield of barley per acre for fifty farms. Using the same principles as before, an arbitrary origin is placed at the group 19–19·5 and arbitrary units allotted to all the groups. Using these arbitrary units the mean yield comes to be $\tfrac{1}{50}(-7)$ or $-0{\cdot}14$.

Table 6.3. *Yield of barley per acre*

(1) Yield (cwt.)	(2) Arbitrary units	(3) No. of farms	(4) Product (2) × (3)	
18–	−2	3	−6	−20
18·5–	−1	14	−14	
19–	0	21		
19·5–	1	11	11	13
20–	2	1	2	
	Total	50		
			Final total	−7

It is now necessary to relate this mean to the original units, First the arbitrary origin of zero corresponds to a yield of 19·25 cwt. and hence a mean of zero in the arbitrary units would correspond to 19·25 cwt. Similarly a mean of -1 corresponds to 18·75, or $19{\cdot}25 - 1 \times 0{\cdot}5$, and a mean of -2 corresponds to 18·25, or $19{\cdot}25 - 2 \times 0{\cdot}5$. Thus by simple proportion a mean of $-0{\cdot}14$ corresponds to $19{\cdot}25 - 0{\cdot}14 \times 0{\cdot}5$ or 19·18 cwt., which is therefore the required mean.

The whole procedure can be expressed as a rule. Let h be the group interval (h was equal to 1 for example 6.1 and 0·5 for example 6.2). Then the mean in the original units is

$$\text{arbitrary origin} + (h \times \text{mean in arbitrary units}).$$

6.6 The grouping adopted for any particular set of data is bound to introduce, as has been stated, some form of error or approximation. In practice it is found, however, that provided the procedure outlined above for the selection of groups is followed, the resulting error will be insignificant. For example, in the heights of schoolboys the true mean was 62·635 in., whereas the mean calculated from the grouped distribution was 62·62. The error introduced by the grouping is therefore $-0{\cdot}015$ in. As the heights were originally measured only to the nearest tenth of an inch, the error can be seen to be trifling compared with the accuracy of the measurement. If the group interval were smaller, and consequently more groups, the error could be expected to be even less, but if the group interval were increased the error might also increase. Hence the general conclusion is that, provided the groups are not too wide, no appreciable error will be introduced by calculating a mean from the grouped distributions instead of from the original observations.

If the original observations can take only a definite set of values such as the integers (0, 1, 2, 3 and so on) then the method used for grouped data above can be directly applied to calculate the mean. No grouping takes place so no grouping errors are committed. This is the situation in the next example.

Table 6.4. *Distribution of persons per household*

No. of persons (k)	1	2	3	4	5	6	7	8	9	10	Total
No. of households with k persons in them	26	113	120	95	60	42	21	14	5	4	500

Example 6.3 Table 6.4 gives the number of persons that were found in 500 households. It is required to find the mean number of persons per household. The mean number of persons per household will be the total number of persons found, divided by the number of households in which they are contained. Now

$$\text{Total number of persons} = 1 \times 26 + 2 \times 113 + \ldots + 10 \times 4 = 1{,}888$$

since 26 households have just one person, 113 households have exactly two persons and so on. Hence the mean will be

$$\tfrac{1}{500}(1{,}888) = 3{\cdot}776.$$

This is an exact result. There are no errors due to the grouping of the data as there can be no values in between the integers.

6.7 Although the mean is by far the most commonly used measure of central tendency of a distribution and will often recur in the later parts of this book, two other measures of average value are sometimes used and should be understood. The first of these is the *median*. This is defined as a value, x, such that half the observations are greater than x and half the observations are smaller than x. For instance in the case of the height of schoolboys it is required to find some height x such that 50 boys have heights greater than x and 50 boys have heights less than x. This could be found by trial and error. Taking the original table of heights, table 3.7, it is found that

26 boys have heights below 61·5 in.

48 boys have heights below 62·5 in.

68 boys have heights below 63·5 in.

from which it can be deduced that the median is just above 62·5 in. If a count is made of the number of boys with heights below 62·6 in. that is, up to and including 62·5 in., there are 49 of them. Hence if the heights of the boys are arranged in ascending order of magnitude, the 49th boy will have a height of 62·5 in. and the 50th and 51st boy will both have a height of 62·6 in. Hence the median must lie somewhere between 62·6 (the height of the 50th boy) and 62·6 (the height of the 51st boy)! Thus the median in this instance is taken as being 62·6 in. In general if there is an even number of observations, as in this case, the median is taken as the average of the two central values, which may or may not be the same. In this particular instance, with a hundred observations, the median is taken as the average of the 50th and 51st observation.

If the number of observations is odd the median is taken as the middle value. For instance, if there were 101 observations the median value would be the 51st in order of magnitude, as there would then be 50 observations both above and below the value. Thus the median value of the five Bristol heights in section 6.1 would be 61·6 since 59·1 and 58·3 are below this value, and 62·1 and 64·7 above it.

6.8 There is another method by which the median may be found which is of especial use in large distributions where the location of the middles value can be very tedious. To use this method it is first necessary to construct a *cumulative frequency distribution*. This is done in table 6.5 (using the data of heights from table 3.10). The table gives for each height the number of boys whose height is less than or equal to that value.

Table 6.5. *Cumulative distribution of heights*

Height (in.)	No. of boys	Height (in.)	No. of boys
Up to 54·95	1	Up to 62·95	58
Up to 55·95	2	Up to 63·95	74
Up to 56·95	2	Up to 64·95	86
Up to 57·95	3	Up to 65·95	93
Up to 58·95	5	Up to 66·95	97
Up to 59·95	9	Up to 67·95	99
Up to 60·95	18	Up to 68·95	100
Up to 61·95	36		

A graph can be plotted of this table. Height is put along the horizontal axis and numbers of boys from 0 to 100 along its vertical axis. Points are then plotted corresponding to the heights and numbers of boys in table 6.5. These points are then joined by straight lines, giving the cumulative distribution diagram shown in fig. 6.2.

This diagram provides a very useful means of obtaining various values that do not appear directly in table 6.5. Thus from the diagram it can be estimated that the number of boys whose height is less than 64·5 in. is 80. This is obtained by reading off the

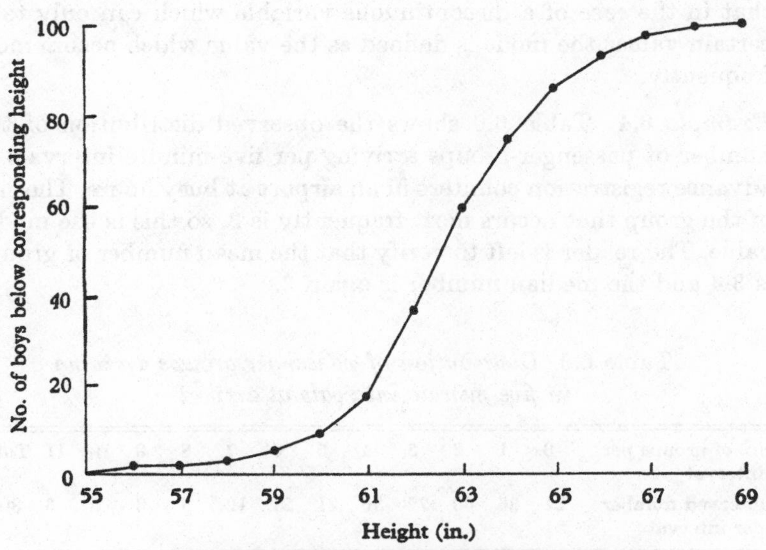

Fig. 6.2. Cumulative distribution of heights

diagram the ordinate of the point on the graph corresponding to 64·5 in. along the horizontal axis.

For the median a value is required that cuts the observations in half and is, therefore, between the 50th and 51st values. For all practical purposes the median can be taken as the value corresponding to 50·5 on the vertical scale, that is $\frac{1}{2}(50+51)$, thus necessitating finding only one value. In this case, using a somewhat larger scale diagram than that shown, the median then comes to 62·6 in. This graphical method should be applied only for grouped frequency distributions in which it is difficult to determine the central value by any direct method; for example, when the group intervals are all different.

6.9 The third and last measure of central tendency to be considered is the *mode*. As the word itself implies the mode is the most frequent or most 'fashionable' value. To take a simple illustration, postage stamps are available in various denominations from a halfpenny upwards, but the fourpenny stamps, being those required for ordinary second class inland letters, are the most frequently used. Hence in the distribution of sales of postage stamps by denominations it is the modal value. From this it follows that in the case of a discontinuous variable which can only take certain values the mode is defined as the value which occurs most frequently.

Example 6.4 Table 6.6 shows the observed distribution of the number of passenger-groups arriving per five-minute intervals at advance registration counters in an airport at busy hours. The size of the group that occurs most frequently is 3, so this is the modal value. The reader is left to verify that the mean number of groups is 3·4 and the median number is again 3.

Table 6.6. *Distribution of passenger-groups arriving in five-minute intervals at airport*

No. of groups per interval	0	1	2	3	4	5	6	7	8	9	10	11	Total
Observed number per interval	21	36	60	72	36	21	21	12	9	6	3	3	300

The determination of the mode of a continuous variable such as height raises a more difficult problem, because if the measurements of height were made with sufficient accuracy, it might well be that no two of the measurements would be the same. To give the central value of the group containing the greatest frequency can be misleading, as this will depend on the choice of scale for the group intervals. If the intervals are made smaller to avoid this difficulty the frequencies in the groups will then become small and the distribution irregular, thus making it difficult to locate the mode. However, these difficulties can be overcome if it is remembered what was said in chapter 5 on frequency curves. The mode, being the most common value, corresponds to the peak or highest point of the frequency curve. Bearing this in mind, a reasonable guess as to the mode can be made by using the grouped frequency distribution to visualise the underlying frequency curve. Thus from the distribution of heights in table 3.9 it can be seen that the mode occurs somewhere in the group going from 62 to 63 in. As the groups on either side of this one contain about equal numbers the mode probably lies at about the middle of the group, that is at about 62·5 in. It is impossible in such cases to determine the mode any more accurately.

6.10 It is interesting to compare geometrically the three measures used here to locate the distributions. For the symmetrical distribution of fig. 6.3 the three measures will all coincide at the central value. The central value must be the mean as every value above it is counterbalanced by an equivalent one below. As it is also the value that cuts the distribution in half, it must indicate the

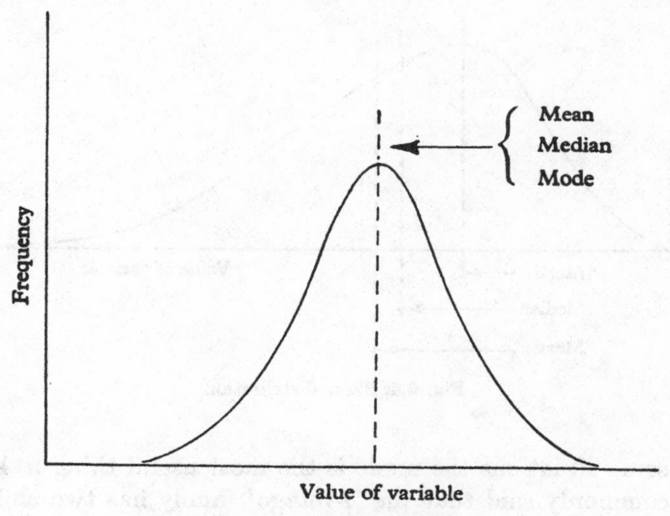

Fig. 6.3. Symmetrical distribution

median, and being the most common value it is thus the mode as well. When there is a skew form of frequency curve such as in fig. 6.4 the three measures do not coincide any longer. This illustrates how necessary it is to use the same measure when comparing two distributions; if this is not done like will not be compared with like.

Of the three measures of central value or location discussed above, the one that will be used almost exclusively in this book is the mean. It is the only one of the three measures which makes complete use of all the observations and hence it can be expected to be more representative of the whole distribution than the other two measures. The median and the mode are often, however, found with much less work, and there are occasions when they are more use than the mean. For instance, it sometimes happens that the exact values of a few extreme observations are not known, and in such a case the mode and median can still be found, although the mean cannot be evaluated. In exercise 4.9, for example, the mean ages of males and females cannot be calculated, but the two medians could be found and compared.

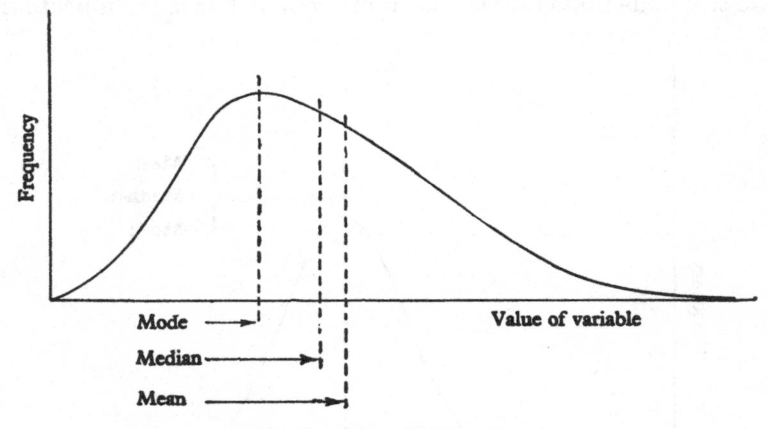

Fig. 6.4. Skew distribution

In some situations the mode is the most useful thing to know. It is commonly said that the 'average' family has two children. This figure is not the mean number of children per family. In fact the mean does not turn out to be an exact integer and a mean family is thus impossible to imagine. The statement implies that more families have two children than any other number of children, and hence it is the modal number of children per family. This fact is relevant in the building of houses for it is obviously important to meet the needs of the most common size of family and for this purpose the mode is required. But again the mode by itself is not enough, since some families have one or three children and so different sizes of house are required. Thus it is necessary to know whether all the observations are concentrated at the mode or whether there is some spread of the observations around the mode. This requires a knowledge of the dispersion of the observations and will be considered in the next chapter.

EXERCISES

The means, medians and modes of the various distributions studied in earlier chapters should all be calculated wherever possible. The exercises suggested below by no means exhaust the material that is available from the earlier chapters.

6.1 Using the data of exercise 4.4 calculate the mean number and the modal number of deaths recorded per day.

6.2 From the data of exercise 4.6 calculate the mean number of children passing G.C.E. at O-level and at A-level per year.

6.3 Using the data of exercises 4.14 and 4.15 calculate the means and medians of the four quantities involved. Carry this out
(*a*) from the original measurements;
(*b*) using an arbitrary origin;
(*c*) by calculation from a grouped frequency distribution.

6.4 From the data collected in exercise 2.23 find the mean length of the piece of line marked off.

6.5 Use the data collected in exercise 2.25 to find the mean and modal length of sentence for the various authors studied.

6.6 The table gives the diastolic blood pressure of 250 men. The readings were made to the nearest millimetre and the central value of each group is given.

Blood pressure (mm.)	No. of men	Blood pressure (mm.)	No. of men
60	4	80	114
65	5	85	30
70	31	90	25
75	39	95	2

Calculate from the data (*a*) the mean, (*b*) the median; and (*c*) make an estimate of the mode of the distribution.

6.7 The following table gives the weight of 1,000 men to the nearest pound:

Weight (lb.)	No. of men	Weight (lb.)	No. of men
Under 100	6	180–199	212
100–119	43	200–219	144
120–139	93	220–239	40
140–159	191	Over 239	8
160–179	263	Total	1,000

(*a*) Calculate the median.
(*b*) Estimate the mode of the distribution.

(*c*) It is impossible to calculate the mean directly from the data but by making reasonable assumptions for the end-groups calculate the mean and compare it with the median and mode.

6.8 In a factory a lathe has been set to produce 1 in. screws. The following table shows the lengths of 800 screws, measured to the nearest one-thousandth of an inch.

Length (in.)	No. of screws	Length (in.)	No. of screws
0·993	3	0·999	94
0·994	26	1·000	19
0·995	85	1·001	8
0·996	160	1·002	5
0·997	230	1·003	1
0·998	169	Total	800

Calculate the three types of average for this distribution and compare the results.

6.9 The following table from the *New Survey of London Life and Labour* (1934), gives the family size of 484 working-class families in Kensington in 1929.

No. of persons	1	2	3	4	5	6	7	8 and over	Total
No. of families	70	112	104	84	43	41	18	12	484

(*a*) Calculate the mode and median of the distribution.

(*b*) By making reasonable assumptions, suggest limits within which the mean will almost certainly lie and compare these limits with the results obtained under (*a*).

6.10 Calculate the mean, median and mode for the ages of American railroad male employees who were members of a retirement scheme in 1944 (data due to R. J. Myers).

Age (years)	No. of men (thousands)	Age (years)	No. of men (thousands)
10–14	1	45–49	294
15–19	289	50–54	265
20–24	225	55–59	233
25–29	261	60–64	150
30–34	283	65–69	62
35–39	284	70–74	13
40–44	324	Total	2,674

6.11 Calculate the mean, median and mode for the number of bracts on specimens of wild carrot collected in Michigan (data due to W. D. Baten).

No. of bracts	4	5	6	7	8	9	10	11	12	13	14	15	16	Total
No. of specimens	1	7	8	41	303	224	140	127	93	52	1	2	1	1,000

6.12 The table below, taken from W. G. Cochran, gives the number of patients who drop out of a clinic roster in two-month periods. Calculate the median time that a patient remains on the roster.

No. of months	No. dropping out	No. of months	No. dropping out
0–	61	12–	12
2–	23	14–	11
4–	14	16–	6
6–	13	18–	5
8–	17	20–	4
10–	14	Over 22	52

6.13 The shoulder widths of fifty-six three-month-old infants are tabulated below. Calculate the mean, median and mode for the distribution (data due to N. Bayley and F. C. Davis).

Shoulder width (cm.)	No. of infants	Shoulder width (cm).	No. of infants
14·4	1	16·5	10
14·7	1	16·8	3
15·0	3	17·1	6
15·3	2	17·4	2
15·6	4	17·7	4
15·9	5	Total	56
16·2	15		

6.14 A group of 5,000 drivers sustained the numbers of accidents given below. Calculate the mean and median for this distribution. Why does the mean not coincide with either the mode or the median?

No. of accidents	0	1	2	3	4	5	6	7	Over 7	Total
No. of drivers	3,140	1,202	423	155	50	15	5	3	7	5,000

6.15 Estimate the mean and median of the following distribution of sales per tobacconist shop, stating any assumptions you may make. The data are taken from the Census of Distribution.

Sales (£)	No. of tobacconists
Under 1,000	1,149
1,000–2,500	1,009
2,500–5,000	1,177
5,000–10,000	2,434
10,000–25,000	3,469
25,000–50,000	885
Over 50,000	161

6.16 The following distribution shows the time taken to service passengers at a registration counter at an airport. Estimate the mean and median times per passenger. Comment on the reason for the difference between the two figures.

Time (min.)	No. of passengers
0–	15
1–	35
2–	25
3–	8
4–	6
5–	4
6–	3
7–	2
8–	1
9–	1
10–	0

7

MEASURES OF DISPERSION

7.1 The last chapter described in some detail the concept of an average and how it can be calculated. But, as was stated there, the average by itself is not sufficient to describe a set of data completely, nor to make valid comparisons between two sets of data. Two illustrations will make these points clear.

A politician states that the average weekly wage in industry X is £16.36 in comparison with a minimum weekly wage of £14 in industry as a whole. This, however, may or may not be indicative of a satisfactory state of affairs. If all the workers earn £16.36 then all are above the mininum of £14. If, however, 60 per cent of the workers earn £13 per week and the remaining 40 per cent earn £20.14 per week, then the average or mean wage is still at £16.36 even though 60 per cent of the workers earn less than the reasonable minimum wage.

Again, imagine two cricketers A and B. Both cricketers have an average batting score of 50, but whereas in ten innings cricketer A achieves scores of

0, 112, 97, 4, 76, 1, 88, 102, 14, 6,

the scores of cricketer B were

42, 71, 51, 39, 60, 44, 58, 47, 51, 37.

Thus cricketer A seems to make either a very low or a very high score every time that he goes in to bat and since he does each about an equal number of times his average comes out to be 50. On the other hand cricketer B appears to be a much steadier player. He rarely makes a duck or a century but nearly always gets a score in the range 35–65 and his final average is once again 50. Here, then, are two cricketers with the same average, but the information about their respective scoring abilities which is given by their averages alone is misleading, as it gives the impression that the two cricketers are the same. In fact cricketer A is not nearly as reliable a batsman as the rather steady cricketer B. To overcome this difficulty use must be made of some measurement that summarises the spread of the distribution of the scores. The scores made by cricketer A have a much larger spread than those of cricketer B and thus any measure proposed should bring out this distinction. As in the case of central tendency there are several possible measures, which will be described in turn.

7.2 The first measure is called the *inter-quartile distance* and its calculation is somewhat analogous to the procedure used in the calculation of the median. The median was defined as the value such that 50 per cent of the observations were above and 50 per cent below it. Suppose now that each of these two halves of the observations are further split into two equal parts. There are now four parts to the distribution and each contains 25 per cent of the observations. Then the value below which just 25 per cent of the observations fall is called the lower quartile, Q_1 say, and the value above which 25 per cent of the observations fall is called the upper quartile, Q_3. Clearly 50 per cent of the observations fall between Q_1 and Q_3. Then the inter-quartile distance (I.Q.D.) is defined as

$$\text{I.Q.D.} = Q_3 - Q_1$$

and is the distance between the upper and lower quartiles. This quantity is very easy to calculate in practice and has a simple meaning, as it is the spread that will contain the central half of the observations. This system of formulation also ensures that the measure is not upset by any extreme or freak observations that could not be called representative values. The possible methods of calculation closely resemble those used for the median.

Example 7.1 Table 7.1 gives the number of accidents sustained by 166 lorry drivers over a long period of time. To calculate the inter-quartile distance it is necessary to find Q_1 and Q_3. 25 per cent of the total number of observations, 166, is $41\frac{1}{2}$ so that Q_1 could be taken as being half-way between the 41st and 42nd observations when arranged in order of magnitude, and Q_3 as being half-way between the 124th and 125th observations in order of magnitude. By counting, or by forming a cumulative distribution from the data, it is found that, when arranged in order of magnitude, the

41st observation is 5 124th observation is 10

42nd observation is 5 125th observation is 10

Hence $Q_1 = 5$ $Q_3 = 10$

and the inter-quartile distance is $Q_3 - Q_1 = 5$.

Table 7.1. *Accidents sustained by drivers*

No. of accidents	No. of drivers	No. of accidents	No. of drivers
0	1	11	9
1	2	12	6
2	3	13	2
3	14	14	6
4	17	15	1
5	21	16	6
6	17	17	3
7	14	19	2
8	14	21	3
9	12	Total	166
10	13		

Example 7.2 To find the inter-quartile distance for the observations of height given in table 3.7.

These heights were plotted in the form of a cumulative frequency distribution in fig. 6.2 and the graph was then used to estimate the 50 per cent value for the median. The same graph can now be used to estimate the values of Q_1 and Q_3 by finding the height corresponding to the 25 per cent and 75 per cent points. From the graph it is possible to estimate that

$$Q_1 = 61{\cdot}35 \quad \text{and} \quad Q_3 = 64{\cdot}05.$$

Hence $$Q_3 - Q_1 = 2{\cdot}70 \text{ in.}$$

The inter-quartile distance is very useful when some of the values are not exactly specified. For example, in a distribution of incomes the last group may be merely labelled 'incomes over £10,000' so that it is difficult to give any numerical value to the individuals in that group for the purposes of calculation. In the inter-quartile distance, however, it is not usually necessary to know the exact values in the end groups. A disadvantage of the I.Q.D. is that it does not make use of every observation, and a variety of rather different-looking distributions could produce the same value of inter-quartile distance. For this reason the other two measures to be considered make use of the numerical values of all the observations.

7.3 The next measure to be discussed is the *mean deviation*. The calculation of this is best described in stages by using an example.

Example 7.3 Consider the sets of cricket scores for cricketers A and B given in section 7.1. The mean score for each cricketer, obtained by summing the scores and dividing by the number of innings, is 50. Next subtract the mean score from each individual score giving

Cricketer A $-50, 62, 47, -46, 26, -49, 38, 52, -36, -44,$

Cricketer B $-8, 21, 1, -11, 10, -6, 8, -3, 1, -13.$

It will be noticed that for each cricketer the sum of the ten deviations from the mean is zero. This must be the case since each of the original scores can be regarded as being composed of two parts, the average plus the deviation from the average, and as the sum of all the scores is ten times the average, the sum of the deviations from the average is always zero.

Now since the sum of all these deviations is automatically zero, a measure of spread cannot be based on them as they stand. If the magnitude of the deviations is used, irrespective of their sign, a measure is obtained which is a guide to the spread of the observations. This is called the mean deviation and is obtained by summing the deviations regardless of their sign and dividing by the number of observations involved. Thus:

Mean deviation for cricketer A

$$= (50+62+47+46+26+49+38+52+36+44)/10$$

$$= \tfrac{450}{10} = 45.$$

Mean deviation for cricketer B

$$= \tfrac{82}{10} = 8{\cdot}2.$$

The mean deviation for cricketer A is, therefore, some five times that of cricketer B. This was expected as the spread of the scores was very much greater for A than it was for B.

When the data are available in the form of a grouped frequency table the necessary calculations have to be slightly modified.

Example 7.4 Suppose that it is required to find the mean deviation of the heights of boys from the data of table 3.10, reproduced in columns (1) and (2) of table 7.2. The mean of this distribution has already been calculated and was equal to 62·63. Column (3) then shows the deviation of the centre of each group from the mean of

62·63. All the deviations are given a positive sign. The group containing the mean is omitted from the calculations at this stage and will be dealt with later. Column (4) gives the contribution of each group to the total deviation from the mean and is the number of observations multiplied by the corresponding deviation, and hence is equal to (2) × (3). This column is summed to give 167·92. To this figure must be added the contribution to the total deviation from the group 62·45 which contains the mean. If the whole frequency is assumed to be concentrated at the mid-point of the group, the contributions to the deviation from these observations at the top and bottom of the group would not be given their correct weight and in the limiting case, when the mean coincides with the mid-point, the whole group would contribute, erroneously nothing to the total deviation.

Table 7.2. *Calculation of mean deviation*

(1) Height (in.) central values	(2) No. of schoolboys	(3) Deviation from mean	(4) Product (2) × (3)
54·45	1	8·18	8·18
55·45	1	7·18	7·18
56·45	0	6·18	0·00
57·45	1	5·18	5·18
58·45	2	4·18	8·36
59·45	4	3·18	12·72
60·45	9	2·18	19·62
61·45	18	1·18	21·24
62·45	22	*	*
63·45	16	0·82	13·12
64·45	12	1·82	21·84
65·45	7	2·82	19·74
66·45	4	3·82	15·28
67·45	2	4·82	9·64
68·45	1	5·82	5·82
Total	100		167·92

Assuming, as a reasonable approximation, that the observations within the group are equally spaced there will be

$22 \times 0{\cdot}68$ observations below the mean in the group

and $22 \times 0{\cdot}32$ observations above the mean in the group.

Whereas the former will have an average deviation of $\frac{1}{2} \times 0{\cdot}68$ from the mean, the latter will have an average deviation of

$\frac{1}{2} \times 0{\cdot}32$. Hence the total deviation from the mean of the observations in the group will be

$$22 \times 0{\cdot}68 \times \tfrac{1}{2} \times 0{\cdot}68 + 22 \times 0{\cdot}32 \times \tfrac{1}{2} \times 0{\cdot}32$$
$$= 11[(0{\cdot}68)^2 + (0{\cdot}32)^2] = 6{\cdot}21.$$

This must now be added to the total deviation already found, giving a final total deviation of 174·13 and (dividing by the total number of observations, 100) a mean deviation of 1·74 in.

If the original observations are available it is feasible to calculate the mean deviation using each individual by itself. The values of 'observation minus mean of observations' are formed and then summed, disregarding their signs. This sum, divided by the number of observations, gives the mean deviation.

7.4 The third measure of dispersion to be defined is the *standard deviation*. The computation of this measure depends to a large extent on the form in which the data are available, and will be illustrated by three examples that give the original data in rather different forms.

Example 7.5 Refer once more to the two cricketers of section 7.1 and write down the deviations of each cricketers' scores from the mean as before. Underneath each deviation write its square.

Cricketer A										
Deviations	−50	62	47	−46	26	−49	38	52	−36	−44
(Deviations)^2	2,500	3,844	2,209	2,116	676	2,401	1,444	2,704	1,296	1,936
Cricketer B										
Deviations	−8	21	1	−11	10	−6	8	−3	1	−13
(Deviations)^2	64	441	1	121	100	36	64	9	1	169

The sum of these squared deviations is now found for each cricketer:

Cricketer A: 21,126, Cricketer B: 1,006,

then the average value of these squared deviations will be obtained by dividing each of the total deviations by 10, giving

Cricketer A: 2,112·6, Cricketer B: 100·6

as the average squared deviation of the observations from the mean value. Since all the deviations were squared at the start it is convenient to take the square root of the average sum of squares, so that the final measure has the same dimensions as the original observations from which it was obtained. Thus

$$\text{Cricketer } A\text{: } \sqrt{2{,}112{\cdot}6} = 46{\cdot}0,$$

$$\text{Cricketer } B\text{: } \sqrt{100{\cdot}6} = 10{\cdot}0,$$

and these are the *standard deviations* of the two distributions of scores. As in the mean deviation the dispersion of cricketer B is about one-fifth of the dispersion of cricketer A. It will also be noticed that the values of the mean deviation and standard deviation are about the same. Usually the mean deviation is smaller than the standard deviation and is something like 80 per cent of it. This is not a rule but just an approximate guide which often helps to check calculations. In this case the differences between the two measures are smaller, due to the rather peculiar U-shaped form of distribution that cricketer A has for his scores.

7.5 If there is a large number of observations the process of calculating the squared deviations can become very tedious. Some short cuts are therefore used. Their form depends on whether the variable concerned has a discrete or a continuous form of distribution. The next example deals with a discrete form of distribution.

Table 7.3. *Calculation of standard deviation of number of defective components*

(1) No. of defective components	(2) No. of cartons	(3) Arbitrary units	(4) (2) × (3)		(5) (3) × (4)
0	2	−3	−6	−46	18
1	8	−2	−16		32
2	24	−1	−24		24
3	36	0	0		0
4	19	1	19	74	19
5	12	2	24		48
6	5	3	15		45
7	4	4	16		64
Totals	110	—	Final total	28	250

Example 7.6 This example uses the data concerning the number of defective components found in 110 cartons, each carton containing a gross of components. The first four columns in table 7.3

are exactly as they would be for the calculation of the mean, which is first calculated giving

$$\frac{74-46}{110} = 0{\cdot}2545$$

in the arbitrary units. To calculate the standard deviation this mean could now be subtracted from each of the central values, the values squared and the squared deviations added up. This would involve a lot of work, especially if there were a large number of groups, and to avoid this an alternative method of formulation for the standard deviation is used. This states:

(Standard deviation)^2 = Average of the squared individual values minus (mean value)^2.

The statement is algebraically equivalent to the previous statement:

(Standard deviation)^2 = Average value of (individual value minus mean)^2 for all the observations.

From the second definition it can be seen that the standard deviation is unaltered if a constant amount is added to any individual as the mean is increased by the constant amount as well. It follows that the standard deviation can be calculated using an arbitrary origin and the result would be the same as if the original units were used. Thus the necessary procedure for this example is to square the central values of the arbitrary units, multiply by the number of observations corresponding to the central value and sum for all the groups. Column (5) gives the appropriate values for each group since (3) × (4) is equal to (2) × {(3)}^{2}. The sum of this column is 250. Hence

$$\begin{aligned}(\text{Standard deviation})^2 &= \tfrac{1}{110} \times 250-(0{\cdot}2545)^2\\ &= 2{\cdot}2727-0{\cdot}0648 = 2{\cdot}2079,\end{aligned}$$

and standard deviation = 1·486 components. No correction is required for the use of an arbitrary origin but it must be remembered that arbitrary units are used throughout. For example, the mean that is used for subtraction is the mean in the arbitrary units and is not converted back to the original units.

7.6 Before describing the third case some easy algebraic notation will be introduced to make the argument simpler to follow. Let x represent an individual measurement such as a cricketer's score or the height of a schoolboy. Let $\bar{x}$ represent the mean of a number of individuals and let s represent the standard deviation of the individuals. Now the deviation of one individual from the mean is

$$x-\bar{x},$$

and if this is squared it becomes $(x-\bar{x})^2$. These squared deviations are now summed, giving

$$\Sigma(x-\bar{x})^2,$$

where Σ is a symbol that represents the statement 'add up all the values of...'. Thus $\bar{x} = \frac{1}{n}\Sigma x$ if there are n individuals. Finally

$$s^2 = \frac{1}{n}\Sigma(x-\bar{x})^2, \tag{7.1}$$

s^2 being referred to as the *variance*. This was the formula and argument used for example 7.5. In example 7.6 a slightly modified form of formula was used where

$$s^2 = \frac{1}{n}\Sigma n_x . x^2 - \bar{x}^2, \tag{7.2}$$

and n_x is the number of observations in a group for which the variable has the value x. Since any arbitrary origin may be used, a judicious choice of origin can often considerably reduce the arithmetic involved. Equation (7.2) is the formula most commonly used for the calculation of s, and, by taking $n_x = 1$, can be used with ungrouped data where each individual forms a group of one, giving

$$s^2 = \frac{1}{n}\Sigma x^2 - \bar{x}^2. \tag{7.3}$$

When calculating the mean from a grouped distribution no correction was required, as it was expected that any errors committed by assuming the individuals to be concentrated at the middle point would cancel out. In the calculation of the standard deviation from the grouped form of continuous distribution a slight correction is required. This arises because the assumption that all the individuals are concentrated at the central value of the group gives, in general, an over-estimate of the standard deviation. Hence instead of (7.2) the formula is

$$s^2 = \frac{1}{n}\Sigma n_x . x^2 - \bar{x}^2 - \tfrac{1}{12}h^2, \tag{7.4}$$

where h is the width of the groups of the distribution. For example, h is 1 in. in the case of the schoolboys' heights in table 3.10. The square root of expression (7.4) gives the standard deviation as before and the correction due to group width will usually be very small.

7.7 *Example* 7.7 The data in table 7.4 give the carbon content, expressed as a percentage, of 119 scoops of a powder. The central values of the groups are given in column (1) while the number of scoops whose carbon content falls in each group is given in column (2). Column (3) gives the arbitrary units. The origin is taken somewhere in the centre of the distribution and the groups labelled $+1$, $+2$, and so on in one direction and -1, -2 and so on in the other direction. Column (4) gives the values of $x.n_x$, where x is the value of the group and n_x the number of individuals in that group. Column (5), equal to (3) × (4), gives the values of $x^2.n_x$.

Table 7.4. *Calculation of standard deviation of carbon content*

(1) Percentage carbon (central value)	(2) No. of samples in group	(3) Arbitrary units	(4) (2) × (3)		(5) (3) × (4)
4·645	1	−4	−4	−64	16
4·745	5	−3	−15		45
4·845	14	−2	−28		56
4·945	17	−1	−17		17
5·045	29	0	—		—
5·145	22	1	22	108	22
5·245	15	2	30		60
5·345	9	3	27		81
5·445	6	4	24		96
5·545	1	5	5		25
Totals	119	—	Final total	44	418

Now $$\Sigma x.n_x = -64+108 = 44$$
is obtained by adding up column (4). Similarly by adding up column (5)
$$\Sigma x^2.n_x = 418.$$

Hence $$\text{Mean} = \tfrac{1}{119}\Sigma x.n_x = \tfrac{44}{119} = 0·3697$$

and $$\frac{1}{n}\Sigma x^2.n_x = \tfrac{1}{119}.418 = 3·5126.$$

Carbon content is a continuous variable that has been grouped and hence equation (7.4) is the appropriate definition of standard deviation. The value of h is equal to 1 because the calculation is being carried out in the arbitrary units and for these arbitrary units the group interval is unity. Hence

$$s^2 = 3{\cdot}5126-(0{\cdot}3697)^2-0{\cdot}0833$$
$$= 3{\cdot}2926 \qquad \text{and } s = 1{\cdot}8146,$$

thus giving the standard deviation in the arbitrary units as 1·8146. The result must now be converted to the original units. From table 7.4 it can be seen that a change of 1 in the arbitrary units corresponds to a change of 0·1 in the original units of percentage carbon. Thus a spread of 1·8146 in the arbitrary units will correspond to a spread of 0·18146 in the original units. This gives the standard deviation of the original distribution as 0·1815 per cent.

Next, convert the mean calculated in the arbitrary units back to the original units. Zero on the arbitrary scale corresponds to 5·045 on the original scale and a change of 1 corresponds to 0·1 in the original units. Hence the mean in the original units is

$$5{\cdot}045+(0{\cdot}3697)(0{\cdot}1) \quad \text{or} \quad 5{\cdot}082.$$

In this calculation arbitrary units have been used throughout and only at the end have the results been converted to the original units of the distribution. This system is adopted in order to avoid mistakes which may otherwise occur if arbitrary and original units are mixed.

7.8 There is one refinement of the standard deviation and variance that will be dealt with more formally later, but needs to be mentioned at this stage. In the definition of variance given in (7.1) the divisor was n, the number of observations. If the observations are only a sample (or sub-set) of a larger set of observations and the variance calculated from this sample is to be used to estimate the variance of the larger set, then the divisor is $n-1$ and not n. In other words the appropriate formula becomes

$$s_1^2 = \frac{1}{n-1}\Sigma(x-\bar{x})^2. \tag{7.5}$$

The estimated standard deviation is then s_1. The change is usually a marginal one numerically, but may be important if the sample (sub-set) is small, i.e. n is small.

The effect can be illustrated on example 7.7. Using formulae (7.5) and (7.2) the formula can be modified to read

$$(n-1)\, s_1^2 = \Sigma n_x . x^2 - n\bar{x}^2. \tag{7.6}$$

For example 7.7 this formula gives

$$(118)\, s_1^2 = 418-(119)\,(\tfrac{44}{119})^2 \quad \text{or} \quad s_1^2 = 3{\cdot}4046.$$

Next the adjustment for grouping, as in formula (7.4) must be applied to give an adjusted value of

$$s_1^2 = 3{\cdot}4046 - 0{\cdot}0833 = 3{\cdot}3213 \quad \text{and} \quad s_1 = 1{\cdot}8224$$

which gives a standard deviation of 0·1822 on the original units. Notice that the difference between the two calculations is only of the order of ½ per cent, and the difference would be even less if there were more observations.

7.9 In the succeeding chapters of this book the standard deviation will be the measure of dispersion that is invariably used. This comes about because it is much easier to handle mathematically than either of the other measures that have been suggested. The mean deviation, with its necessary sorting out of negative and positive deviations, is somewhat unwieldy, whilst the inter-quartile distance does not possess the simple algebraic properties of the standard deviation. Nevertheless there are cases in which the inter-quartile distance may be the appropriate quantity to calculate; for example, when the end groups do not have definite boundaries.

If the variability in two sets of data is to be compared it is important that the same measure is used in two cases. If this is not done there is a danger that apparent differences are due, not to the variables being measured, but to the different measures of dispersion being used.

In making a comparison of, say, two standard deviations, it is essential that they are measured in the same units as the original observations. Thus the standard deviation of the height of school-boys in a particular case is 2·12 in. but this could equally well be expressed as 0·1767 ft. Further it is impossible to compare with any validity the standard deviations of two variables that have different basic units, for example, height and weight. This must be so for, whilst height could be measured in inches or centimetres say, weight could be measured in pounds or kilo-grams, and which units should be used for the comparison? Different answers could be obtained by using different sets of units.

If it is desired to discover whether one distribution is relatively more variable than another, it follows that it is necessary to find some method of eliminating the basic units. This is achieved by using the *coefficient of variation*, defined thus:

$$\text{Coefficient of variation} = \frac{\text{Standard deviation}}{\text{Mean}} \times 100 \text{ per cent.} \quad (7.7)$$

This coefficient does not depend on the units of measurement since both the mean and standard deviation are linear functions of the units involved. If the unit of measurement is changed from pounds to kilo-grams by multiplying every observation by the factor 0·4536, both the standard deviation and the mean will be multiplied by the same factor and hence the coefficient of variation is unaltered.

The coefficient of variation for the carbon content in a powder in example 7.7 will be

$$\frac{s}{\bar{x}} \times 100 = \frac{0{\cdot}1815}{5{\cdot}082} \times 100 = 3{\cdot}57 \text{ per cent,}$$

and this would be unaltered if all the original observations were multiplied by any factor.

7.10 Measures of dispersion are as important characteristics of a series of observations as were the measures of position. It is vital that in any form of inquiry attention should be directed not solely to the average value of the observations but also to the manner in which those observations are distributed about the average. So far in this book the methods available for the collection of data, its reduction to tables and charts and some basic descriptive measures have been dealt with. Such treatment of a problem would enable broad general conclusions to be drawn and decisions made whenever a very large volume of data was involved. The next stage must be to examine the situation when the volume of data is not large, and to see how such measures as the mean and standard deviation would fluctuate if only a selection of the possible observations were available. By such a study it is possible to see how representative of the whole field are the available observations. This basic and fundamental problem of statistics will now be considered at some length. It is, however, essential to understand fully the methods of presentation of data and the calculation of measures of central value and of dispersion. These are basic processes which will constantly be needed in the material that follows.

EXERCISES

The previous chapters contain a very large number of distributions for which the inter-quartile distance, the mean deviation and the standard deviation can be calculated. The exercises suggested here do not exhaust the material available for such calculations.

7.1 Calculate the inter-quartile distance for the tables in exercises
(*a*) 4.17 Deaths of persons in railway accidents.
(*b*) 4.9 Ages of males and females in 1947. Treat each sex separately.
(*c*) 6.7 Weights of men.
(*d*) 6.9 Family size in Kensington in 1929.

7.2 Calculate the mean deviation for the tables in exercises
(*a*) 4.4 Deaths recorded per day in *The Times*.
(*b*) 4.11 Degree of cloudiness at Greenwich.
(*c*) 3.4 Weights of pigs.
(*d*) 3.5 Yield of mangold roots.
(*e*) 4.15 Span of 60 men. (Use the table formed in the exercise for the calculation.)
(*f*) 6.8 Length of screws.

7.3 Calculate the standard deviation for the tables in exercises
(*a*) 3·1 Telephone calls.
(*b*) 3.2 Words per sentence. (Use the raw data as given.)
(*c*) 4.16 Tensile strengths of malleable iron castings.
(*d*) 6.6 Diastolic blood pressure.
(*e*) 4.14 Heart weight.
(*f*) 6.10 Age of railroad employees.

7.4 Calculate the mean deviation and standard deviation for the following sets of data and calculate also the ratio standard deviation/mean deviation. Does this ratio vary very much from distribution to distribution, or does it remain approximately constant?
(*a*) 4.15 Forearm lengths. (Use the table formed in the exercise for the calculation.)
(*b*) 3.10 Tensile strength and hardness. (Use table formed in the exercise.)
(*c*) 3.7 Age at onset of tuberculosis. (Use table formed in the exercise.)

7.5 Calculate the mean deviation and the standard deviation of the following table which gives the lengths of 237 specimens of the fruit of the blood-root (*Sanguinaria canadensis*).

Length (mm.)	No. of specimens	Length (mm.)	No. of specimens
24–27	1	45–48	32
27–30	1	48–51	22
30–33	10	51–54	5
33–36	25	54–57	4
36–39	39	57–60	2
39–42	41	60–63	—
42–45	54	63–66	1
		Total	237

7.6 The following table gives the distribution by age of 996 miners in South Africa suffering from miner's phthisis. Calculate the mean, the mean deviation and the standard deviation of the miners' ages. Note carefully the method of classification that has been used.

Age (years)	No. of miners	Age (years)	No. of miners
15–19	9	45–49	75
20–24	56	50–54	47
25–29	192	55–59	13
30–34	239	60–64	4
35–39	217	65–69	3
40–44	140	70–74	1
		Total	996

7.7 The data below give the distance, in centimetres, that the top of the head is above the ear, for 235 schoolgirls aged 10 years. Calculate the mean and standard deviation for the data.

Height (central values)	No. of girls	Height (central values)	No. of girls
10·5	1	12·3	34
10·7	—	12·5	33
10·9	4	12·7	21
11·1	3	12·9	17
11·3	8	13·1	7
11·5	14	13·3	8
11·7	23	13·5	4
11·9	25	13·7	2
12·1	29	13·9	2
		Total	235

7.8 The table gives the weight, in ounces, of the zinc coating on 75 sheets of galvanized iron. Calculate the mean and standard deviation of the data.

Weight of coating (oz.) (central values)	1·30	1·35	1·40	1·45	1·50	1·55	1·60	1·65	1·70	1·75
No. of sheets	1	5	6	13	8	17	14	7	1	3

7.9 Calculate from the following data the mean and standard deviation of the age of onset of the eye disease, optic atrophy, in Japanese males.

Age at onset (years)	No. of males	Age at onset (years)	No. of males
4–7	1	28–31	8
8–11	2	32–35	4
12–15	20	36–39	1
16–19	20	40–43	3
20–23	11	44–47	—
24–27	3	48–51	1
		Total	74

7.10 Calculate the standard deviation of the number of noxious weed seeds in ninety-eight quarter-ounce packets of *Phleum pratense* seeds.

No. of noxious weed seeds	0	1	2	3	4	5	6	7	8	9	Total
No. of packets	3	17	26	16	18	9	3	5	—	1	98

7.11 Calculate the mean and standard deviation of the milk yield of cows given in the table. The data refer to one year.

Milk yield per cow (gallons)	No. of cows	Milk yield per cow (gallons)	No. of cows
200–299	2	700–799	41
300–399	10	800–899	22
400–499	23	900–999	6
500–599	54	1000–1099	1
600–699	61	Total	220

7.12 The monthly sales of an American department store over two years are given below in thousands of dollars.

834, 866, 892, 898, 950, 957, 1054, 1094, 1095, 1097, 1099, 1099, 1101, 1102, 1103, 1105, 1107, 1111, 1121, 1122, 1151, 1171, 1175, 1206.

Calculate the mean and standard deviation of the values (*a*) directly from the figures given, (*b*) by forming a grouped distribution with an interval of 50 and using the grouped data.

7.13 The table below gives the number of earners found per family in a social survey that was carried out in a large town. Calculate the mean and standard deviation.

No. of earners in family	No. of families	No. of earners in family	No. of families
0	117	5	47
1	1086	6	14
2	616	7	4
3	287	8	1
4	126	Total	2,298

7.14 The table below gives the central value weights, in grams, at birth of 310 twin babies of the same sex. Calculate the standard deviation and the coefficient of variation for this data (due to M. Fraccaro).

Weight (gm.)	No. of babies	Weight (gm.)	No. of babies
500·5	2	2600·5	68
800·5	10	2900·5	50
1100·5	4	3200·5	20
1400·5	10	3500·5	7
1700·5	21	3800·5	2
2000·5	54	Total	310
2300·5	62		

7.15 Calculate the mean deviation and the standard deviation of the inner diameter measurements of the cylinders given below. What is the ratio of the mean deviation to the standard deviation? Can you explain why the ratio is not 0·8 as it usually is?

Diameter (0·0001 in.)	No. of cylinders	Diameter (0·0001 in.)	No. of cylinders
90	1	101	2
91	2	102	1
92	—	103	3
93	1	104	2
94	—	105	5
95	—	106	2
96	—	107	1
97	2	108	—
98	2	109	—
99	2	110	2
100	2	Total	30

(Data adapted from J. R. Crawford.)

8

PROBABILITY AND SAMPLING

8.1 An ordinary penny was tossed ten times and the result of the tossings was seven heads, three tails. If nothing else were known about the behaviour of tossed coins except this one experiment, its result would have to be used to estimate the frequency with which a coin comes down heads when tossed. The proportion of heads occurring in the set of ten tossings was $\frac{7}{10}$, or 0·7, and this would be the best estimate from the experimental results. When the whole procedure was repeated and a further ten tossings made the result was five heads, five tails. From this second experiment the estimated proportion of heads was $\frac{5}{10}$, or 0·5, which is less than before. In all twenty tossings, however, there were twelve heads so that a better estimate of the proportion of heads in repeated tossings would be $\frac{12}{20}$, or 0·6. When this simple experiment of ten tossings was repeated a large number of times, the numbers of heads in the first ten sets were respectively:

7, 5, 5, 6, 4, 5, 4, 5, 6, 6.

The proportion of heads at any stage is the total number of heads observed divided by the total number of tossings then performed. After ten tossings the proportion of heads is $\frac{7}{10}$, or 0·7, whilst after twenty tossings it is $(7+5)/20$, or 0·6, and the successive results are:

Total no. of tossings	10	20	30	40	50	60	70	80	90	100
Proportion of heads	0·70	0·60	0·57	0·58	0·54	0·53	0·51	0·51	0·51	0·53

The proportion of heads fluctuates a great deal while the number of tossings is small, but as the experiment proceeds the fluctuations get damped down and become small. If more tossings are made the fluctuations get even smaller, so that the proportion of heads observed would appear to tend to a limit. Thus the proportion of heads after 100 tossings was 0·53, after 200 tossings it was 0·52, after 300 tossings it was 0·52, after 500 tossings it was 0·51, and after 1000 tossings it was still 0·51 to two decimal places.

For investigations of the proportionate frequency with which an event occurs it is not essential that there are just two possible cases, such as head and tail, each of which appears about an equal number of times. An ordinary die has six faces and the number of sixes obtained in ten tossings was noted. In this case the two categories are 'six' and 'not six' and it is soon apparent that the two alternatives do not appear in anything like equal proportions. If it is a perfectly constructed die and the method of tossings does not favour any one particular face, a long series of tossings would show that the proportion of sixes, although subject to large fluctuations at first, settles down to about $\frac{1}{6}$.

The essential feature to recognise in these experiments is that in a short-term, or small-scale, experiment, the proportionate frequency of some event may not be the same as it would be in a large-scale experiment. If the proportionate frequency is worked out continuously as the number of experiments increases, the fluctuations in the proportionate frequency become less and less noticeable and eventually the proportion will remain stable from one experiment to the next.

8.2 In order to be able to use the proportionate frequency of events obtained from such experiments the theory of *probability* is needed. The word probability has been defined in several ways but the simplest definition is to state that probability is 'the proportionate frequency of occasions on which some stated event occurs'. For the prediction of future events the proportionate frequency required will be that of the limiting case, when the number of occasions or experiments carried out is large. Thus in section 8.1 the probability of a tossed coin giving heads is the proportion of times that heads occur in repeated tossings of the coin. After a thousand tossings this proportion is 0·51 and in an infinite set of tossings of which the thousand form part, it seems likely that the proportion would be exactly 0·5. From the definition, probability must lie somewhere on a scale from zero to one. Zero corresponds to absolute impossibility—that you could fly unaided—whilst one corresponds to absolute certainty—that you will die some day. In between there is a range of possibilities, some events falling near one end of the scale (winning the £25,000 prize on the premium bonds would fall near the zero end) whilst some fall near the other end of the scale. For example, when a doctor says that penicillin will clear up your complaint he is not really implying that this is absolutely certain, only that it is very nearly certain. Finally, tossing a fair coin gives a 50:50 chance of a head implying that it is midway between these two extremes of impossibility and certainty.

8.3 Most problems in probability are concerned with the happening not of one event only but of two or more events. The possible results of interest are referred to as the basic outcomes and the set of all possible basic outcomes is referred to as the *sample space*. Suppose that employees of a firm are classified first by sex and secondly according to whether they are under 35 years of age or not. Then fig. 8.1 gives a symbolic representation (sometimes

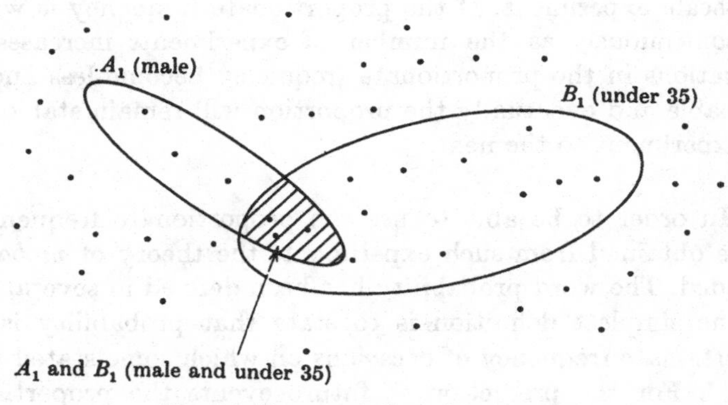

Fig. 8.1. Representation of two events

referred to as a Venn diagram) of the possibilities. The area A_1 represents those who are male, the remaining area represents those who are female. The area B_1 represents those who are under 35 years of age, the remaining area represents those who are 35 or over. The shaded area of overlap represents those who satisfy both A_1 and B_1, i.e. they are male and under 35.

Events are said to be *mutually exclusive* if they cannot occur together. For example, if the two events were that a die when tossed showed six on its uppermost face and two on its lowest face, these events would be mutually exclusive, as the ordinary die has one on the face opposite six. On the other hand if the two events were that the uppermost face had a one on it and the lowest face had a number greater than three, the events could occur together in one tossing and are not mutually exclusive. Such a situation is illustrated by the Venn diagram shown in fig. 8.2.

Using this definition the first basic theorem of probability can be stated.

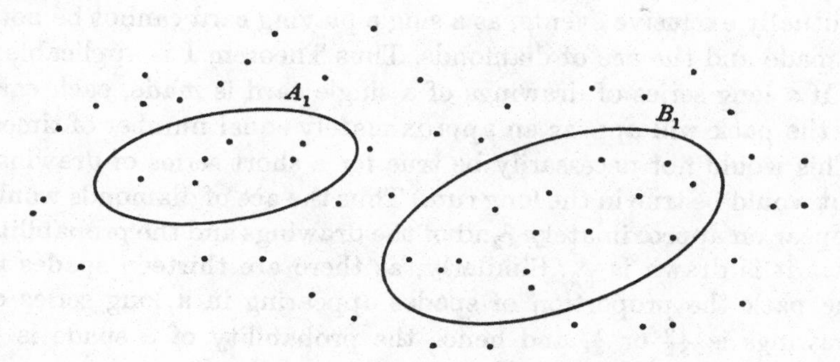

Fig. 8.2. Mutually exclusive events

Theorem I. If A and B are two events then the probability of either A or B occurring is equal to the sum of the probability that A occurs with the probability that B occurs minus the probability that both A and B occur. Symbolically this is written as

$$\Pr\{A+B\} = \Pr\{A\}+\Pr\{B\}-\Pr\{AB\},$$

where $\Pr\{A\}$ stands for the probability of event A occurring, $\Pr\{AB\}$ stands for the probability of both A and B occurring and $\Pr\{A+B\}$ stands for the probability of either A or B (or both) occurring. The logic of this theorem can be seen from a study of fig. 8.1. If the two events are mutually exclusive, then $\Pr\{AB\}$ must be zero, as shown by fig. 8.2. Hence the theorem reduces to

$$\Pr\{A+B\} = \Pr\{A\}+\Pr\{B\}.$$

Example 8.1 Alpha and Omega are two weather stations in Wales. Let A and W represent the occurrence of rain at Alpha and Omega respectively, during a 24-hour period in August. It was found that $\Pr\{A\} = \Pr\{W\} = 0{\cdot}4$ and $\Pr\{AW\} = 0{\cdot}28$. What is the probability of rain at either Alpha or Omega (or both)?

$$\begin{aligned}\Pr\{A+W\} &= \Pr\{A\}+\Pr\{W\}-\Pr\{AW\}\\ &= 0{\cdot}4+0{\cdot}4-0{\cdot}28\\ &= 0{\cdot}52.\end{aligned}$$

Example 8.2 A card is drawn at random from an ordinary pack of 52 playing cards. It is required to find the probability that the card drawn is either a spade or the ace of diamonds. Let A be the event 'spade' and B be the event 'ace of diamonds'. They are mutually exclusive events, as a single playing card cannot be both a spade and the ace of diamonds. Thus Theorem I is applicable.

If a long series of drawings of a single card is made, each card in the pack will appear an approximately equal number of times. (This would not necessarily be true for a short series of drawings but would be true in the long run.) Thus the ace of diamonds would appear on approximately $\frac{1}{52}$nd of the drawings and the probability that it is drawn is $\frac{1}{52}$. Similarly, as there are thirteen spades in the pack the proportion of spades appearing in a long series of drawings is $\frac{13}{52}$ or $\frac{1}{4}$, and hence the probability of a spade is $\frac{1}{4}$. This gives

$$\Pr\{A\} = \tfrac{1}{4}, \qquad \Pr\{B\} = \tfrac{1}{52}.$$

Hence

$$\Pr\{A+B\} = \Pr\{A\}+\Pr\{B\}$$

$$= \tfrac{1}{4}+\tfrac{1}{52} = \tfrac{14}{52} = \tfrac{7}{26}.$$

Example 8.3 A man has ten coins in his pocket, two half-crowns, three florins, one shilling and four sixpences. He draws out a coin at random from his pocket. What is the probability that it is either a shilling or a sixpence?

Let event A be that the coin is a shilling whilst event B is that the coin is a sixpence. These are again mutually exclusive events. If repeated drawings of a single coin from a man's pocket were made (the coin being replaced after each drawing), each coin would appear an equal number of times. It follows that the probability of drawing a coin of any particular denomination is the proportion of coins of that denomination. Hence the probabilities of the events A and B are

$$\Pr\{A\} = \tfrac{1}{10}, \qquad \Pr\{B\} = \tfrac{4}{10},$$

giving

$$\Pr\{A+B\} = \tfrac{1}{10}+\tfrac{4}{10} = \tfrac{1}{2}.$$

It should be noted that this theorem only expresses symbolically a very simple argument, namely that since five coins out of the ten satisfy the required condition the expected probability is $\frac{5}{10}$ or $\frac{1}{2}$.

8.4 *Independence.* Two events are said to be independent if the happening of one event does not affect the happening of the other. For example, suppose that the experiment consists of drawing one card from a pack of playing cards and the first event is that the card is a knave and the second that it is a heart. If a long series of drawings were made and the occasions when a knave is drawn noted, it would be found that the four knaves in the pack had occurred about the same number of times. This confirms that the four suits occur equally often amongst the knaves and, hence, one knave in four is a heart. This proportion is exactly the same as the proportion of hearts in single cards drawn from the whole pack. Hence the two events are said to be independent. On the other hand, suppose that the experiment consisted of selecting a card at random, the first event being that the card was not a court card and the second event being that the points value of the card was above eight. These two events will not be independent, because $\frac{3}{10}$ of the non-court cards are of the value nine or above, whilst $\frac{6}{13}$ of the whole pack are nine or above in value (ace high). Thus the occurrence of the first event, card not a court card, would affect the occurrence of the second event. Such a pair of events cannot be said to be independent, and are therefore dependent events. This leads to a second theorem.

Theorem II. If A and B are two independent events then the probability that both A and B occur simultaneously is equal to the product of the probabilities that A and B occur separately. Written symbolically the theorem reads

$$\Pr\{AB\} = \Pr\{A\} \times \Pr\{B\},$$

where $\Pr\{AB\}$ stands for the probability of both event A and event B occurring.

Example 8.4 From a pack of playing cards two cards are drawn at random the first card being placed before the second is drawn. What is the probability that the first card drawn is a heart and the second an ace?

A is the event that the card drawn is a heart. From previous examples the probability of a heart being drawn is taken to be the proportion of hearts in the pack and therefore

$$\Pr\{A\} = \tfrac{13}{52} = \tfrac{1}{4}.$$

Similarly for event B there are four aces in the pack of fifty-two cards and the proportion of aces is $\frac{1}{13}$. Hence

$$\Pr\{B\} = \tfrac{1}{13}.$$

Since the first card is replaced before the second card is drawn, the two drawings, and hence the two events, are independent of one another, as the happening of one event does not affect the happening of the other. Theorem II can then be applied giving

$$\begin{aligned}\Pr\{AB\} &= \Pr\{A\} \times \Pr\{B\} \\ &= \tfrac{1}{4} \times \tfrac{1}{13} = \tfrac{1}{52}.\end{aligned}$$

Example 8.5 Two dice are thrown and it is known that for each die all the faces are equally likely to come uppermost. What is the probability that the total score thrown is two?

For the total score to be equal to two both dice must show a one on the uppermost face. Let A be the event that the first die shows a one and B the event that the second die shows a one. Then $\Pr\{A\} = \Pr\{B\} = \frac{1}{6}$, and, since the tossings of each die are independent, the probability of the joint event that both show a one is

$$\Pr\{AB\} = \Pr\{A\} \times \Pr\{B\} = \tfrac{1}{36}.$$

8.5 The next two examples make use of both theorems stated above. All readers are advised to work through these examples carefully.

Example 8.6 Bag I contains four white and four black balls whilst bag II contains one white and seven black balls. A bag is chosen at random and a ball then chosen at random from the bag. What is the probability that the ball chosen is white?

The experiment consists of two stages, first choosing a bag and then choosing a ball. A white ball can arise in one of two mutually exclusive ways:

1. Bag I is selected and a white ball then drawn.
2. Bag II is selected and a white ball then drawn.

The first way consists of two events, namely bag I is chosen and then a white ball is drawn out. Let A be the event of the drawing of bag I and B the event of drawing a white ball from it. Then

$$\Pr\{A\} = \tfrac{1}{2}, \qquad \Pr\{B\} = \tfrac{1}{2}.$$

Since the two events are independent the probability of both occurring is

$$\Pr\{AB\} = \Pr\{A\} \times \Pr\{B\} = \tfrac{1}{2} \times \tfrac{1}{2} = \tfrac{1}{4}.$$

A similar argument can now be used for the second way of obtaining the white ball. If A' is the event that bag II is selected and B' the event that a white ball is drawn from it, then

$$\Pr\{A'B'\} = \Pr\{A'\} \times \Pr\{B'\} = \tfrac{1}{2} \times \tfrac{1}{8} = \tfrac{1}{16}.$$

Since these two ways are mutually exclusive the total probability of getting a white ball will be

$$\Pr\{AB\} + \Pr\{A'B'\} = \tfrac{1}{4} + \tfrac{1}{16} = \tfrac{5}{16}.$$

Example 8.7 A coin is tossed n times. What is the probability that no two consecutive tossings give the same result, assuming that at each toss the coin is equally likely to come down heads or tails?

If the conditions are to be satisfied then each toss must give the opposite result from the previous toss. There are two cases to be considered, namely $HTHTHT\ldots n$ tossings, and $THTHTH\ldots n$ tossings where H stands for a head and T for a tail.

Each of these cases has the same probability, since the tossings are independent. The two cases are also mutually exclusive. The probability of the first case is

$$\tfrac{1}{2} \times \tfrac{1}{2} \times \tfrac{1}{2} \times \ldots \times \tfrac{1}{2} \quad (n \text{ factors}),$$

and hence the overall probability is

$$2 \times (\tfrac{1}{2})^n = (\tfrac{1}{2})^{n-1}.$$

8.6 The ideas in the preceding sections can now be extended to cover the cases where the characteristic concerned is measurable, and not just of the presence or absence variety such as head or not-head. In an investigation to find the average weight of schoolboys it would be impossible to weigh every single schoolboy, and however large a number are weighed they will be only a selection from those available. The complete set of schoolboys is called the *population* and the smaller set selected for weighing is called the *sample*.

To use the concept of probability for such measurable characteristics care must be taken to distinguish between the sample and the population. The depth of sapwood on telegraph poles was measured on five poles, and only one pole had a depth greater than 3·4 in., that is a proportion of 0·2. Further poles were then examined and after twenty poles had been measured, three poles, or 0·15 of those measured, had a depth greater than 3·4 in. By the time 100 poles had been measured the proportion with depth of sapwood greater than 3·4 in. was 0·14, and when 500 poles had been measured the proportion was 0·16. The fluctuations in the proportion became smaller as more poles were measured and the proportion tends to that of the frequency curve corresponding to a depth of sapwood beyond 3·4 in. Figs. 8.3 and 8.4 illustrate how the increase in sample size will bring the frequency distribution nearer to the hypothetical frequency curve discussed in chapter 5, and hence will ensure that the proportion above 3·4 in. becomes more stable. The limiting frequency curve is what a statistician has in mind when he talks of the 'population' of telegraph poles. The probability of a pole having a depth of sapwood greater than 3·4 in. will be the ratio of the area beneath the curve beyond 3·4 in. to that beneath the whole curve.

The definition of proportionate frequency holds good even when interest is not just centred on individuals beyond a certain value. Fig. 8.5 shows the frequency curve corresponding to the length of antennae of the aphis (green-fly). The probability that the length of antenna of one green-fly drawn at random being between 1·6 and 1·8 mm. will be the proportionate frequency with which these lengths occur and is the ratio of the shaded area to the total area beneath the frequency curve.

8.7 The greater the number of observations, the greater is the available information concerning the population from which the observations come. The next question to be considered is how far it is valid to draw inferences about characteristics of the population or universe of individuals from the smaller set of individuals available in the sample. Suppose that in the investigation into weights of schoolboys the sample was selected by taking all the tall boys. There would then be little hesitation in stating that the results of the weighing would be unrepresentative, as tall boys will tend to weigh more than short boys. Thus the sample must be chosen in a manner that is random and unbiased in the sense that every schoolboy is just as likely to appear in the sample as

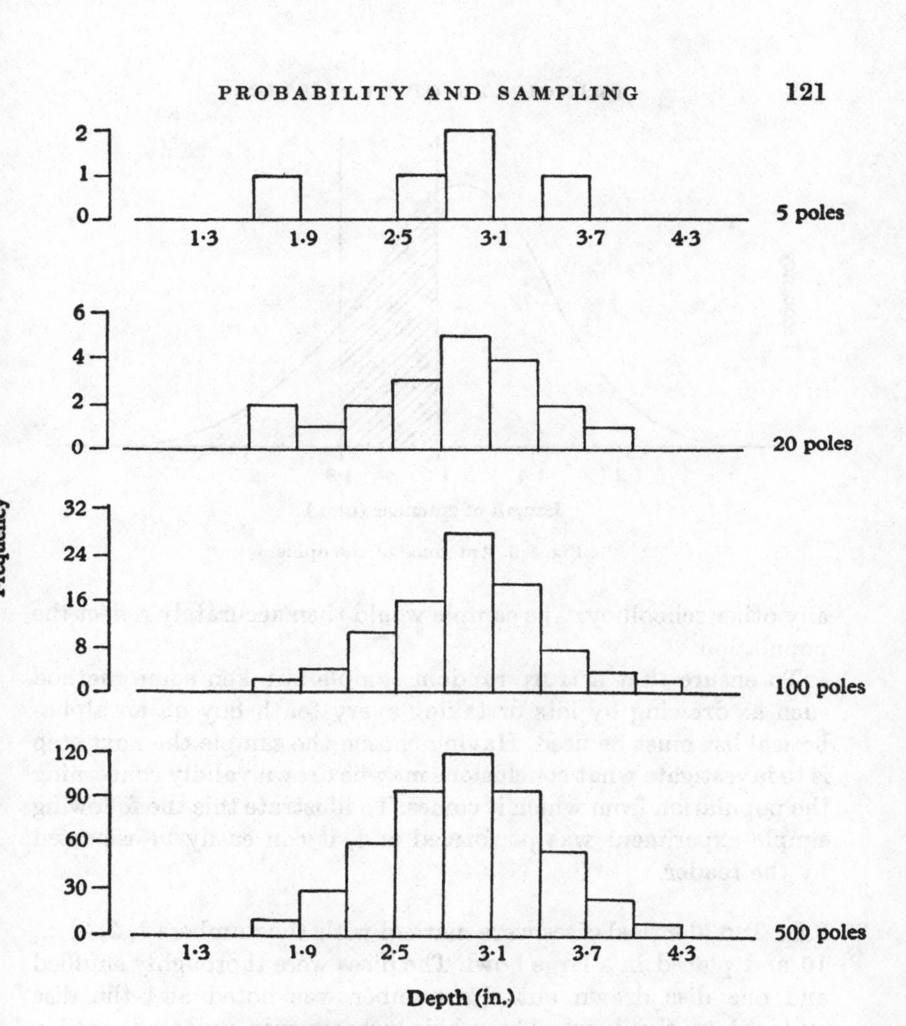

Fig. 8.3. Sapwood depth on telegraph poles

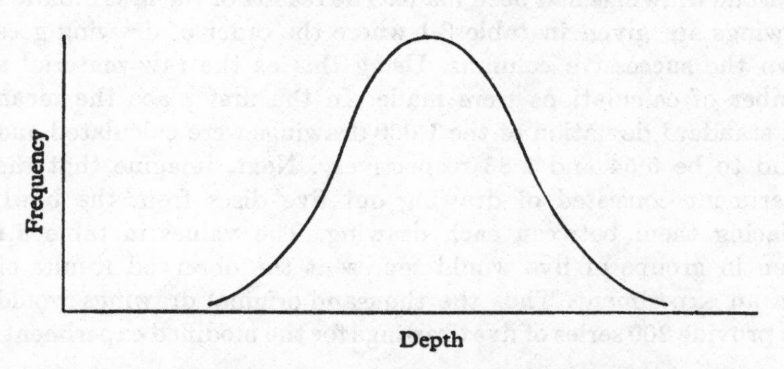

Fig. 8.4. Limiting distribution of sapwood depth on telegraph poles

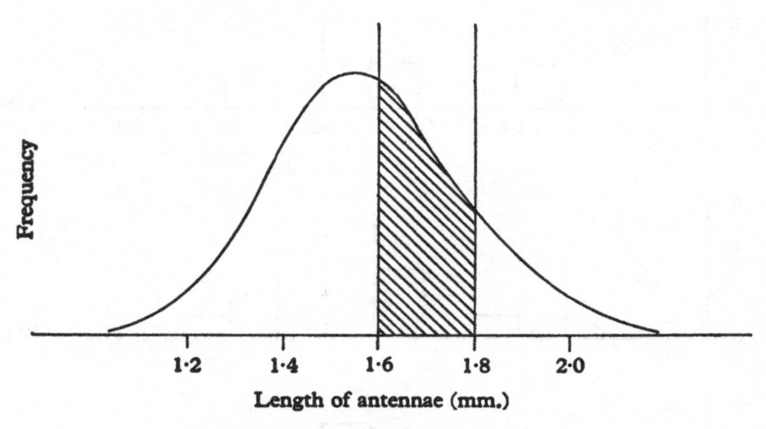

Fig. 8.5. Antennae of the aphis

any other schoolboy. The sample would then accurately reflect the population.

To ensure that a truly random sample is taken some method such as drawing by lots or taking every tenth boy on an alphabetical list must be used. Having chosen the sample the next step is to investigate what conclusions may be drawn validly concerning the population from which it comes. To illustrate this the following simple experiment was performed and, it can easily be repeated by the reader.

8.8 Ten identical discs were marked with the numbers 1, 2, 3, ..., 10 and placed in a large bowl. The discs were thoroughly shuffled and one disc drawn out. Its number was noted and the disc replaced in the bowl. The whole process was repeated until a thousand drawings had been made. The results of the first hundred drawings are given in table 8.1 where the order of drawing goes down the successive columns. Using this as the raw material a number of calculations were made. In the first place the mean and standard deviation of the 1,000 drawings were calculated and found to be 5·54 and 2·83 respectively. Next, imagine that the experiment consisted of drawing out five discs from the bowl, replacing them between each drawing. The values in table 8.1 taken in groups of five would represent the observed results of such an experiment. Thus the thousand original drawings would now provide 200 series of five drawings for the modified experiment.

For each group of five the mean value of the numbers drawn is calculated. For example, the mean of the first set of five is

$$\tfrac{1}{5}(8+3+6+8+9) = 6{\cdot}8.$$

This is done for all the remaining sets of five and the first twenty such means, corresponding to the numbers in table 8.1, are given in table 8.2 reading across the rows. A similar procedure is now carried out for experiments that consist of drawings of groups of 10, 15, 20 and 25 discs and in each case the mean of the groups calculated. These means are now formed into frequency distributions and plotted as dot diagrams in fig. 8.6 where, to avoid overcrowding, not all the dots are given.

Table 8.1. *Drawing of discs*

8	3	1	10	10	3	8	7	3	5
3	6	5	7	7	2	3	2	3	9
6	3	8	3	4	10	9	6	3	3
8	5	2	9	2	4	7	7	8	8
9	1	4	9	2	8	7	2	3	10
8	7	6	10	7	5	2	6	5	8
9	3	2	10	9	6	9	3	10	5
9	4	3	4	7	3	8	7	3	5
1	10	4	2	1	2	4	7	6	10
4	2	5	7	6	5	7	10	1	7

Table 8.2. *Means of five drawings*

6·8	6·2	3·6	5·2	4·0	4·0	7·6	6·6	5·0	6·0
5·4	4·2	6·8	6·0	4·8	6·6	4·0	5·0	7·0	7·0

The most noticeable feature in the figure is that whatever the number of drawings in the experiment the average value of the sample mean is the same. This illustrates the fact that the means of samples are grouped round the mean of the population. A second important feature is that the larger the size of the sample the smaller is the average discrepancy between the sample mean and the population mean. Hence in estimating the population mean, the larger the sample the more accurate will be the estimate.

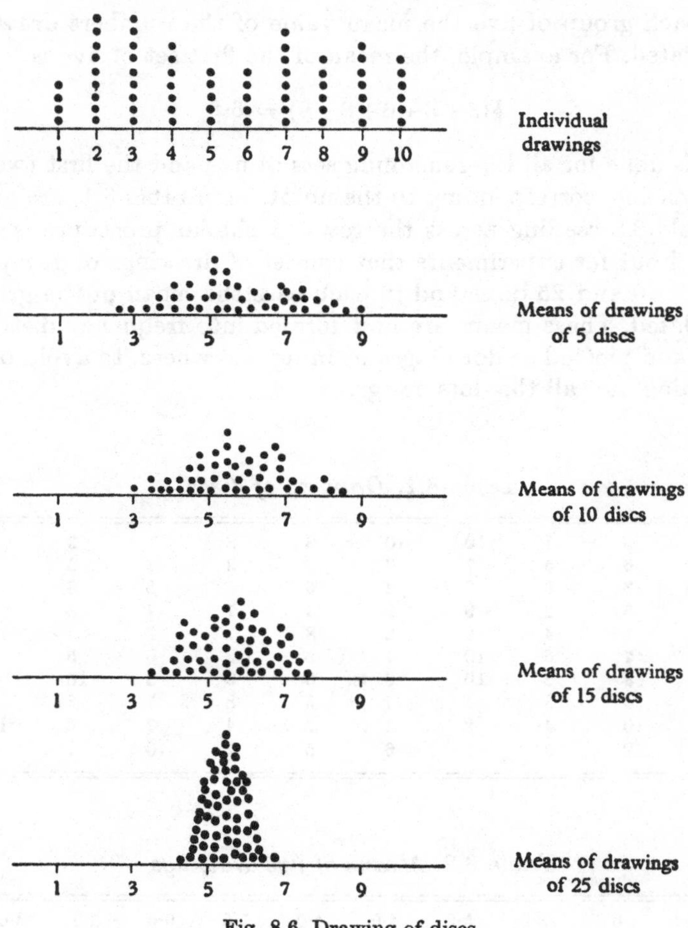

Fig. 8.6. Drawing of discs

8.9 The various distributions just obtained experimentally are termed the 'sampling distributions of the mean', and various calculations were made from the observed results. First the mean and standard deviation of each of the sampling distributions of the mean were calculated. The results are given in table 8.3 and it will be noticed that the mean of the sample means is always the same, as the same 1,000 drawings were used for the experiments. Had the samples been obtained completely afresh for each different sample size there would have been some small variation in these means. The third column shows, by calculating the standard deviation of each distribution, how the scatter of the means about their average value decreases as the size of the sample increases. It is clear that the reduction in standard deviation is not simply inversely proportional to the size of the sample but decreases more slowly than that. For example, when the sample size is doubled, from five to ten, the standard deviation is not halved but is reduced in the ratio of 0·92:1·31 or 1:1·42. The value 1·42 is approximately $\sqrt{2}$ and the general rule for random sampling states that if s is the standard deviation of individuals in a population, then $s/\sqrt{n}$ is the standard deviation of the mean of random samples of size n from that population.

Table 8.3. *Means and standard deviations of sample means*

(1) No. of individuals in the sample (n)	(2) Mean of the sample means	(3) Standard deviation of the means: Observed	(4) Standard deviation of the means: Calculated from $s/\sqrt{n}$
1	5·54	2·83	2·87
5	5·54	1·31	1·28
10	5·54	0·92	0·91
15	5·54	0·71	0·74
20	5·54	0·61	0·64
25	5·54	0·58	0·57

The standard deviation of the ten observations 1, 2, 3, ..., 10 is 2·87 and as each of the ten numbers would appear equally often in a long sequence of drawings this value is taken for s in the population. Column (4) in table 8.3 gives the values of $s/\sqrt{n}$ for the various sizes of samples considered. Comparison of the observed standard deviations with those calculated from $s/\sqrt{n}$ show very close agreement. This empirical property of the variability displayed by the means of random samples of size n from any population is important and forms the basis of much statistical work, since it demonstrates that the larger the sample the more accurate is the sample mean when used for estimating the mean in the whole population.

EXERCISES

8.1 Two ordinary six-sided dice are tossed. What are the probabilities

(*a*) That one die only shows a six?

(*b*) That both dice show the same number?

(*c*) That the sum of the two numbers shown is ten?

8.2 Two players A and B shake a die in turn, A going first. The first player to throw a six starts some game. What are A and B's respective chances of starting the game?

8.3 One of the digits, 1, 2, 3, ..., 7 is chosen at random. What is the chance that the digit will be (*a*) odd, (*b*) even? What do these two probabilities add up to, and why?

8.4 In the Morse alphabet, letters are formed by combinations of dots and dashes. Suppose that all letters had either one, two or three symbols. How many different letters could be formed?

8.5 A bag contains four red balls and three white balls. A ball is drawn out and its colour noted and is not replaced. Another ball is drawn out, colour noted and not replaced, and so on. Calculate the probability that the order of drawing of the balls is red, white, red, white, red, white, red.

8.6 If there are three seals and four colours of sealing wax, in how many ways can a letter be sealed?

8.7 There are three identical pairs of gloves in the hall. A visitor on leaving picks up two gloves at random. What is the probability that the two gloves form a pair if all the selections of gloves are equally likely?

8.8 Two dice are thrown simultaneously. What is the chance that neither gives a one or a six?

8.9 A card is drawn at random from an ordinary pack of playing cards. What are the probabilities of obtaining

(*a*) A heart?

(*b*) The king of hearts?

(*c*) The king or queen of hearts?

(*d*) The king or queen of any suit?

8.10 Three pennies are tossed. What are the probabilities that

(*a*) All three pennies are tails?

(*b*) At least two of the pennies are tails?

8.11 A committee is to be formed of three boys out of six boys eligible to serve on it. In how many different ways can the committee be formed?

8.12 Three men and three women are available for mixed doubles at tennis. How many different games can be made up from the players available?

8.13 There are two urns A and B. Urn A contains three white balls and one black ball; urn B contains two white and three black balls. One ball is now taken from each urn. Calculate the probability that the two balls drawn are one of each colour.

8.14 Three men draw in turn a ball from a bag containing three balls. The balls are identical except that each bears the name of one of the men. Calculate the probability that no man draws a ball bearing his name

(*a*) If the balls are not replaced after each drawing.
(*b*) If the balls are replaced after each drawing.

8.15 A six-sided die is so biased that it is twice as likely to show an even number as an odd number when thrown. It is thrown twice What is the probability that the sum of the two numbers thrown is even?

8.16 The tensile strength of a large number of aluminium die castings has the mean value of 27·1 and a standard deviation of 1·2, the units being 1,000 lb. per sq. in. If samples of size five are taken from this population and the mean tensile strength, x, of the five castings obtained, what will be the mean value and standard deviation of the quantity x in repeated sampling?

8.17 The percentage ash content in a large number of scoops of coal was found to have a mean of 17·92 and a standard deviation of 2·03. Suppose that random samples of n scoops were drawn and the mean ash content in the n scoops found. How large would n have to be for the standard deviation of the mean of n scoops to be less than 0·5?

8.18 The length of the forearm of a very large number of adult males was measured and fell between 17·1 in. and 20·9 in. with a preponderance of values about half-way between the two extremes. If samples of size seven are now drawn from the population of adult males and the mean length of forearm obtained for the sample, between what values, approximately, would you expect the means to lie?

8.19 An electrical mechanism containing four switches will fail to operate unless all of them are closed. The switches are independent with regard to proper closing or failure to close, and for each switch the probability of failure is 0·1. Find the probability of failure of the whole mechanism, ignoring all sources of failure except switches.

9

THE BINOMIAL THEOREM

9.1 This chapter will indicate how a knowledge of the theory of probability as outlined in the previous chapter can be used to make deductions as to the shape of the frequency distributions produced when certain types of experiment are repeated. The classes of experiments considered have two main characteristics in common:

(i) Each experiment is independent of the result of the preceding experiments. Thus the fact that a coin, when tossed, comes down heads does not affect the chance of the coin coming down heads at the succeeding tossing.

(ii) The quantity studied is the presence or absence of some characteristic; that is, there are only two classes to be considered and every event falls into one or other of these. These may be, for example, the heads or tails for a coin tossing experiment, under 6 ft. or over 6 ft. in height for men drawn from some population, or the presence or absence of some defect in articles made by a machine.

In the previous chapter it was found that if repeated independent drawings are made from the population under consideration, the proportion of individuals in the drawings possessing the characteristic concerned will approach the proportion in the whole population possessing it. But in all sets of drawings there will be some variation from the exact proportion, these variations depending on the size of the sets of drawings and the frequency with which the particular characteristic occurs.

9.2 To illustrate how these variations occur a simple experiment was performed. A penny was selected and tossed 2,000 times, the number of heads and tails obtained being recorded. Next two pennies were selected and both tossed. The number of heads obtained, zero, one or two, was noted, and the experiment repeated 2,000 times. The whole procedure was now repeated, using groups of three or four or five pennies, and each time the number of heads was recorded. The results are given in table 9.1.

Several things stand out from this table. First of all the number of times that no heads appear depends very much on n, the number of coins in the group being tossed. For each value of n there were 2,000 experiments and the number of experiments which gave rise to no heads were

1,021, 504, 257, 133 and 71 respectively.

Table 9.1. *Coin-tossing experiment results*

		No. of heads (x)						
		0	1	2	3	4	5	Total
No. of coins in group (n)	1	1,021	979	—	—	—	—	2,000
	2	504	983	513	—	—	—	2,000
	3	257	731	760	252	—	—	2,000
	4	133	518	728	488	133	—	2,000
	5	71	341	609	608	312	59	2,000

It will be noticed that each of the frequencies is about half the preceding frequency. Secondly, the most common number of heads to occur is the central number of those that can possibly occur. Thus if four coins are tossed there are five possibilities, zero to four heads, and the central number would be two heads. From the table this is seen to be easily the most common result. If n is odd there is an even number of possibilities, and the middle two are the most common results. Thus if $n = 3$, one or two heads are the most common results. Finally the approximate symmetry of the table should be noted. For $n = 3$ the number of occasions on which zero or three heads were observed are approximately equal and the number of occasions on which one or two heads were observed are also approximately equal.

Clearly in any particular experiment a table such as table 9.1 could be constructed by a series of experiments under similar conditions. This would be an extremely unwieldy procedure and not of very general use. Table 9.1 has shown that some form of pattern or law emerges, and the next step is to find a general method of deducing the frequencies in the table. This will involve a knowledge of the probability of each individual unit possessing the characteristic concerned.

9.3 Suppose that a coin has a probability p of coming down heads with a single tossing. By this it is meant that in a long series of independent tossings the proportionate frequency with which the coin will come down heads is p. The value of p must, of course, fall between zero and unity. Let $q = 1-p$ represent the probability of a tail. Then if the coin be tossed twice there are four possible outcomes:

(i) first tossing tail, second tossing tail;
(ii) first tossing tail, second tossing head;
(iii) first tossing head, second tossing tail;
(iv) first tossing head, second tossing head.

Written symbolically these four alternatives are

$$TT, \quad TH, \quad HT, \quad HH$$

where H represents a head and T a tail. The two throws are independent, so using Theorem II of chapter 8 the probabilities of the four alternatives are

$$q \times q, \quad q \times p, \quad p \times q, \quad p \times p.$$

The first outcome gives two tails with a probability of q^2. The two middle alternatives each result in one head and one tail and the probability of getting exactly one head is, therefore, by Theorem I of chapter 8, the sum of these two probabilities, as they are mutually exclusive events. Hence the probability is $2qp$. The last alternative gives two heads with a probability p^2. Thus the probabilities of getting 0, 1 or 2 heads in tossing the coin twice are

$$q^2, \quad 2qp, \quad p^2$$

respectively. Note that these statements of probability do not specify the order in which the results occurred but merely the probabilities of overall result. The three probabilities add up to one. If the coin is tossed three times there are now eight possible alternative results, namely

$$TTT, \quad TTH, \quad THT, \quad THH, \quad HTT, \quad HTH, \quad HHT, \quad HHH,$$

and the corresponding alternatives for the eight alternatives are

$$qqq, \quad qqp, \quad qpq, \quad qpp, \quad pqq, \quad pqp, \quad ppq, \quad ppp.$$

The first alternative gives no heads, the second, third and fifth give one head, and the fourth, sixth, and seventh give two heads, whilst the eighth gives three heads. Since the eight alternatives are mutually exclusive the probabilities of the four combined results are:

No heads		q^3,
One head	$pqq+qpq+qqp$	$= 3pq^2$,
Two heads	$ppq+pqp+qpp$	$= 3pq^2$,
Three heads		p^3.

9.4 The method outlined in section 9.3 is quite general and could be extended to any number of tossings of the coin, but the calculation would be long and tedious, and some more general method is obviously required. Now the probability of getting exactly k heads in a simple random sample of n tossings when the order of heads and tails is specified is

$$p^k q^{n-k} \tag{9.1}$$

since there are q tails and p heads. This is merely a generalization of what was deduced in the previous section. Expression (9.1) has now to be multiplied by the number of orders of n trials possible in which exactly k are heads. This is known as the number of combinations that can be formed of k objects out of n and, as shown in algebra text-books, is equal to

$$\binom{n}{k} = \frac{n!}{k!\,(n-k)!},$$

where $x! = x(x-1)(x-2)\ldots3.2.1$ and $\binom{x}{0}$ and $\binom{x}{x}$ are both understood to be equal to one. Hence the required probability of exactly k heads in a series of n independent tossings is

$$\binom{n}{k} p^k q^{n-k}. \tag{9.2}$$

For example, if n is equal to three, the probabilities of 0, 1, 2 or 3 heads in the three trials will be

$$\binom{3}{0} q^3, \quad \binom{3}{1} pq^2, \quad \binom{3}{2} p^2 q, \quad \binom{3}{3} p^3$$

or q^3, $3pq^2$, $3p^2q$, p^3 respectively. It should be noted that the values given by (9.2) for values of k going from 0, 1, ..., n are the successive terms of the expansion of

$$(q+p)^n.$$

A series of expressions giving the probabilities of the various outcomes of an experiment is termed the *probability distribution* of the results. Thus the terms of $(q+p)^3$ form the probability distribution of the number of heads in three independent tossings of a coin.

The expansion $(q+p)^n$ is known as the binomial theorem which states that

$$(q+p)^n = q^n + \binom{n}{1} pq^{n-1} + \binom{n}{2} p^2q^{n-2} + \ldots + \binom{n}{k} p^kq^{n-k} + \ldots + \binom{n}{n-1} p^{n-1}q + p^n.$$

This result is perfectly general and does not merely apply to the tossing of coins. The general statement is that in a series of n independent trials at each of which the probability of some event occurring is constant and equal to p, the probability of exactly k events in the n trials is equal to

$$\binom{n}{k} p^kq^{n-k}, \quad \text{where } q = 1-p.$$

9.5 The calculation of the individual terms is not unduly laborious, although logarithms will sometimes be found useful. Frequently if a series of terms is required it is easiest to proceed in a definite order, with each term building up from the previous one. Thus if

$$P_k = \binom{n}{k} p^kq^{n-k}$$

then the next term is

$$P_{k+1} = \binom{n}{k+1} p^{k+1}q^{n-k-1},$$

and the ratio

$$\frac{P_{k+1}}{P_k} = \frac{n!\,p^{k+1}q^{n-k-1}k!\,(n-k)!}{(k+1)!\,(n-k-1)!\,p^kq^{n-k}n!},$$

which on simplification equals

$$\frac{n-k}{k+1}\frac{p}{q}.$$

Therefore, $P_{k+1} = \dfrac{n-k}{k+1}\dfrac{p}{q} P_k$, an example of a recurrence formula. If P_0 is calculated then the values P_1, P_2, ... can be found in succession.

Example 9.1 A die is thrown four times. Each side of the die has the same probability of $\frac{1}{6}$ of appearing uppermost. If k is the number of sixes that appear in the four throws it is required to find the probability distribution of k.

Table 9.2. *Four throws of a die*

k no. of sixes	P_k algebraic	P_k numerical
0	$(\frac{5}{6})^4$	0·4823
1	$4(\frac{1}{6})(\frac{5}{6})^3$	0·3858
2	$6(\frac{1}{6})^2(\frac{5}{6})^2$	0·1157
3	$4(\frac{1}{6})^3(\frac{5}{6})$	0·0154
4	$(\frac{1}{6})^4$	0·0008

The probability of getting k sixes is $\binom{4}{k}(\frac{1}{6})^k(\frac{5}{6})^{4-k}$, where k takes the values 0, 1, 2, 3 and 4. Arranging the work in tabular form gives the values in table 9.2. The sum of the probabilities is equal to one as it should be, thus demonstrating that one of the five mutually exclusive results given must in fact occur. The results in the table were computed directly but they could equally well be calculated by the recurrence formula. Thus if

$$P_0 = 0{\cdot}4823,$$

$$P_1 = \frac{4-0}{0+1}\times\frac{1/6}{5/6}\times 0{\cdot}4823 = 0{\cdot}3858, \text{etc.}$$

Example 9.2 An experiment is carried out involving the crossing of a fern with a palm. These two varieties of plant can be easily distinguished and the seeds obtained from such a crossing may be planted and the type of plant that comes up noted. According to a theory of cross-fertilisation of plants put forward by Mendel, the chance that the seed resulting from such a crossing should give a palm is 3/4. To test this statement five seeds from a crossing were planted, and the number of palms 0, 1, 2, ..., 5 obtained noted.

Table 9.3. *Observed number of palms obtained*

No. of palms (k)	0	1	2	3	4	5	Total
No. of times k palms were obtained	1	1	5	17	24	12	60

The whole experiment was repeated sixty times, with the results given in table 9.3. If Mendel's theory is correct then the proportion of occasions on which 0, 1, 2, 3, 4 or 5 palms were obtained would be the successive term of

$$(\tfrac{1}{4}+\tfrac{3}{4})^5,$$

since there were five seeds planted and the probability, p, that any one seed would give a palm is 3/4. The terms of this expansion are equal to

$$0{\cdot}0010, \quad 0{\cdot}0146, \quad 0{\cdot}0879, \quad 0{\cdot}2637, \quad 0{\cdot}3955, \quad 0{\cdot}2373$$

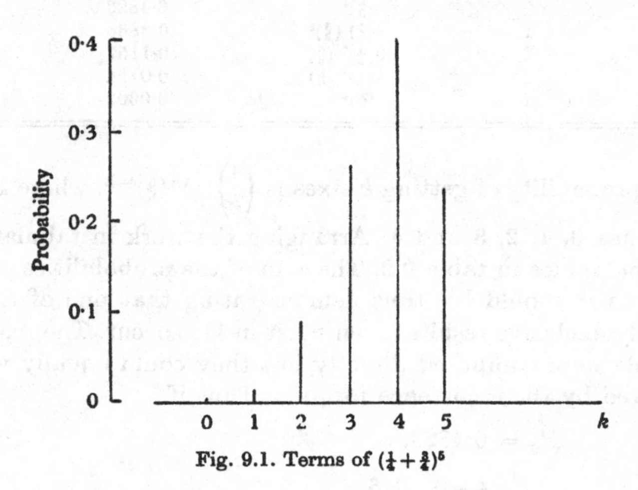

Fig. 9.1. Terms of $(\frac{1}{4}+\frac{3}{4})^5$

and are shown diagrammatically in fig. 9.1 as a frequency diagram. In this experiment sixty crossings were examined. Probability is defined as the proportionate frequency with which some event occurs. Let this probability be P. Then the expected number of occurrences of the event in 60 crossings will be $60P$ since the proportionate frequency of the event concerned is then $60P/60$ or P. Hence the expected frequencies of the six possible outcomes are

$$60\times 0{\cdot}0010, \quad 60\times 0{\cdot}0146, \quad \ldots, \quad 60\times 0{\cdot}2373,$$

or

$$0{\cdot}1, \quad 0{\cdot}9, \quad 5{\cdot}3, \quad 15{\cdot}8, \quad 23{\cdot}7, \quad 14{\cdot}2,$$

which differ little from the observed frequencies shown in table 9.3. In fact the differences seem to be no more than could occur by purely chance fluctuations, and the Mendelian theory would therefore appear to be reasonable on the basis of the observed results of this experiment.

Example 9.3 Consider families containing five children in which there are no twins. Assuming that the probabilities of a child being a boy or a girl are each equal to $\frac{1}{2}$, what fraction of such families could be expected

(i) to have at least one son and one daughter?
(ii) to have all children of the same sex?

If the probability of a child being male is equal to $\frac{1}{2}$ and the sex of each child is independent of the other children, then in families of size five the proportions that have 0, 1, 2, ..., 5 boys will be given by the successive terms of the series

$$(\tfrac{1}{2}+\tfrac{1}{2})^5.$$

On expanding the expression the successive terms are

$$\tfrac{1}{32},\quad \tfrac{5}{32},\quad \tfrac{10}{32},\quad \tfrac{10}{32},\quad \tfrac{5}{32},\quad \tfrac{1}{32}.$$

Now if the family is to have at least one child of each sex amongst the five children it must have between exactly one and exactly four boys. Hence the required proportion will be

$$\begin{aligned}\Pr\{1 \text{ boy}\}+\Pr\{2 \text{ boys}\}+\Pr\{3 \text{ boys}\}+\Pr\{4 \text{ boys}\} &= \tfrac{5}{32}+\tfrac{10}{32}+\tfrac{10}{32}+\tfrac{5}{32}\\ &= \tfrac{15}{16}.\end{aligned}$$

Secondly, if the family is to have all children of one sex it must contain either no boys or else five boys, that is it must be one of the two cases not included in the previous category. Hence the required proportion is

$$\Pr\{0 \text{ boys}\}+\Pr\{5 \text{ boys}\} = \tfrac{1}{32}+\tfrac{1}{32} = \tfrac{1}{16},$$

and it will be noticed that $\frac{15}{16}+\frac{1}{16} = 1$, which is clearly necessary, since the two categories are mutually exclusive and are the only possible categories.

Example 9.4 The probability that a person will respond to a mailed advertisement for a book is 0·2. What is the probability that

(i) two out of a group of ten persons will respond?
(ii) more than two persons out of a group of ten persons will respond?

If the probability of an individual response is 0·2, then the probability of exactly two responses from a group of ten persons is

$$\binom{10}{2}(0{\cdot}2)^2(0{\cdot}8)^8 \quad \text{or} \quad 0{\cdot}3020.$$

For the second part the required probability is

$$\binom{10}{3}(0{\cdot}2)^3(0{\cdot}8)^7 + \binom{10}{4}(0{\cdot}2)^4(0{\cdot}8)^6 + \ldots + (0{\cdot}2)^{10},$$

but it is simpler to calculate this as

$$1-(0{\cdot}8)^{10}-\binom{10}{1}(0{\cdot}2)(0{\cdot}8)^9-\binom{10}{2}(0{\cdot}2)^2(0{\cdot}8)^8$$

which is identical since the sum of the probabilities of the eleven possible outcomes must be unity. Calculation then gives the result as 0·4396.

9.6 The problem may be in a different form. For example, experience may have shown in the past that a certain event occurs with a specified probability. A sample of items is now examined in order to discover whether it is reasonable to assume that the event still occurs with the specified probability or whether the probability has changed.

Example 9.5 Suppose that over a long period of time a manufacturer has been making sparking plugs of which 10 per cent were defective. In an attempt to reduce this percentage, the manufacturer makes a change in his methods. To examine the change a sample of ten plugs is selected at random and tested. There are no defective plugs in the sample. Is it reasonable to assume that the proportion of defective plugs is still 10 per cent?

If the population possesses a proportionate frequency of 0·1 of defective plugs and a random sample of ten plugs is drawn then the probabilities of 0, 1, 2, ..., 10 defective plugs in the sample will be the successive terms of the binomial series

$$(0{\cdot}9+0{\cdot}1)^{10},$$

and the probability of no defective plugs occurring is $(0{\cdot}9)^{10}$, that is the first term, which is equal to 0·349. Thus if the proportion of defective plugs is 0·1 and the experiment of drawing and examining ten plugs is carried out a large number of times, then in about one-third of the experiments it would be found that there were no defective plugs amongst the ten examined. This result, therefore, is quite a common one and would not at all suggest that the figure of 10 per cent of defective plugs had in any way changed.

On the other hand, suppose that in the original sample of ten plugs all ten had been found to be defective. The probability of such a result arising by chance from a population in which 10 per cent are defective is the last term in the binomial expansion given above, namely $(0{\cdot}1)^{10}$ or 0·000,000,000,1. This is very small and such an event should only happen about once in ten thousand million times. If the proportion of defectives in the population of plugs were higher than 0·1 the probability of getting all ten in the sample defective would in turn be much higher, and hence the observed result would throw grave doubts on the belief that the proportion of defectives is 0·1.

Example 9.6 In a certain factory observations over a long period have shown that 20 per cent of the workmen succumb to an occupational disease within a year of commencing work. To try to improve conditions, considerable alterations are made to a particular part of the factory, and of the fifty workmen in this portion only six succumb to the disease during the year following the alterations. Can it reasonably be said that a significant improvement has been effected?

To answer this question consider first the situation if the alterations had effected no improvement. Under these conditions the probabilities of 0, 1, 2, ..., 50 workmen getting the disease in a sample of fifty from a population in which each man has a chance of 0·2 of getting the disease are the successive terms of the binomial expansion

$$(0{\cdot}8+0{\cdot}2)^{50}.$$

Thus $\Pr\{0\} = (0{\cdot}8)^{50}$ and is zero to four decimal places where $\Pr\{0\}$ stands for the probability that none of the men get the disease. Similarly $\Pr\{1\} = 0{\cdot}0002$, $\Pr\{2\} = 0{\cdot}0011$, $\Pr\{3\} = 0{\cdot}0044$, $\Pr\{4\} = 0{\cdot}0128$, $\Pr\{5\} = 0{\cdot}0295$, $\Pr\{6\} = 0{\cdot}0554$.

Hence the probability that exactly six men get the disease in the sample of fifty is equal to 0·0554. It might be decided to use this result to conclude that, as this probability is reasonably small, the true probability of a man getting the disease is smaller than 0·2, because if it were 0·15, say, the observed sample result of six men with the disease becomes a more likely happening. However, it must be borne in mind that if such a decision were to be made when just six of the men in the sample get the disease, an identical decision would be made, only much more strongly so, if five or four or less of the men get the disease since

$$\Pr\{5\} = 0{\cdot}0295 \quad \text{and} \quad \Pr\{4\} = 0{\cdot}0128,$$

and both probabilities are even less than Pr{6}. Hence the decision to say that there is an improvement if only six men get the disease implies that the same decision would be made if five or even less got the disease. The probability that one of these possibilities occurs in the sample when there is no improvement is therefore

$$\Pr\{0\}+\Pr\{1\}+\Pr\{2\}+\Pr\{3\}+\Pr\{4\}+\Pr\{5\}+\Pr\{6\} = 0{\cdot}1034.$$

Thus in about 10 per cent of cases such a result could occur by chance, and this is probably a sufficiently common occurrence for there to be doubt as to whether a significant improvement has in fact been effected. On the other hand, suppose that only three of the fifty men got the disease. The probability of three or less of the men getting the disease if there has been no improvement is

$$\Pr\{0\}+\Pr\{1\}+\Pr\{2\}+\Pr\{3\} = 0{\cdot}0057,$$

and such a result would only occur about once in 180 times by chance. This would throw considerable doubt on the theory that 20 per cent of the men get the disease, and would suggest that in fact the true proportion was now somewhat lower, so that some improvement had been made.

It should be noted that the smallness of the individual probabilities does not by itself prove or disprove the statement to be examined. It is essential to include the probabilities of all results which would lead to the same decision. Secondly, it is necessary for the sample result to be more likely on the alternative theory if the original theory is to be rejected. For example, if out of the fifty men considered above twenty got the disease it will be found that

$$\Pr\{20\} = 0{\cdot}0006,$$

if the probability that each man gets the disease is 0·2. The expected number of men getting the disease is np or 10 so that more extreme values than 20 would be 21, 22 and so on. Adding up these probabilities gives

$$\Pr\{20\}+\Pr\{21\}+\Pr\{22\}+\ldots = 0{\cdot}0009,$$

and shows that such an extreme result would only occur about once in a thousand times or even less. But to reject the proposition that each man has a probability of 0·2 of getting the disease in favour of the alternative that the new conditions have made an improvement would be foolish. This follows because if the probability of getting the disease were in fact lower than 0·2 the probability of getting twenty men with the disease would be even smaller and it would be a still more unlikely event than before.

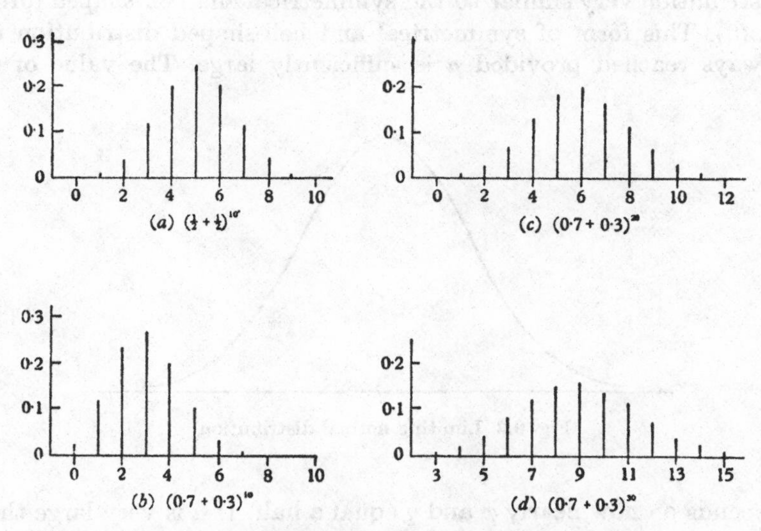

Fig. 9.2. Binomial distributions. (Vertical scale gives probability, horizontal scale number of successes)

9.7 The various examples given demonstrate the need for the calculation of binomial terms, and the last example shows that in many cases it is the sum of a number of end terms that is required. To obtain these values by the separate calculations of each term and then summing up can be very tedious, especially if n is large. Fortunately, however, in such circumstances it is possible to find an easier way of calculating the required probabilities. This is because, as n gets large, the shape of the binomial distributions becomes very similar whatever the particular values of n and p concerned. Fig. 9.2 shows four binomial distributions (*a*) $(\frac{1}{2}+\frac{1}{2})^{10}$, (*b*) $(0·7+0·3)^{10}$, (*c*) $(0·7+0·3)^{20}$, and (*d*) $(0·7+0·3)^{30}$, each plotted in the form of a line diagram. Case (*a*) is symmetrical and bell-shaped because $p = q = \frac{1}{2}$, and it is easy to see that as n increases and the number of ordinates similarly increases the outline will gradually approximate to a smooth symmetrical frequency curve. In case (*b*), however, the distribution is far from symmetrical and is very unlike (*a*) in general characteristics. However, if n is increased from ten to twenty, with p kept constant as it is when going from (*b*) to (*c*), then the distribution becomes more symmetrical, and a further increase of n up to 30, as in (*d*), makes the distribution very similar to the symmetrical and bell-shaped form in (*a*). This form of symmetrical and bell-shaped distribution is always reached provided n is sufficiently large. The value of n

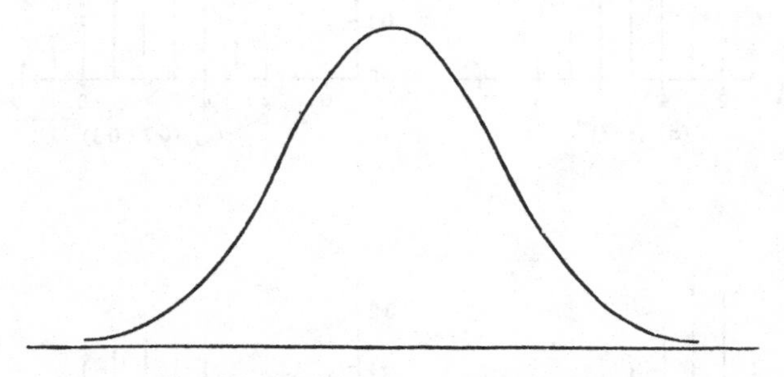

Fig. 9.3. Limiting normal distribution

depends on how nearly p and q equal a half. If n is very large the steps in the line diagram are relatively close together, and when the tops of the lines are joined the diagram approaches the smooth form shown in fig. 9.3 above. This form is called the *normal distribution*, or the Gaussian curve, after the mathematician Gauss (1777–1855). It has a constant shape but the exact location and scale of the distribution depends on two constants, namely, the mean and the standard deviation. In table 9.4 a normal distribution that has a mean equal to zero and a standard deviation of unity is tabulated. Such a distribution is referred to as a unit normal distribution. The area for any given x, shown shaded in fig. 9.4, goes from zero to unity as x increases from minus to plus infinity. The distribution is also symmetrical. Thus the area to the left of the abscissa $-1{\cdot}4$ is $0{\cdot}0808$ whilst the area to the right of $+1{\cdot}4$ is $1-0{\cdot}9192$ or $0{\cdot}0808$. The area on either side of the ordinate at $x = 0$ is clearly equal to $\frac{1}{2}$.

Table 9.4. *Normal curve areas*

Abscissa (x)	Area $F(x)$	Abscissa (x)	Area $F(x)$	Abscissa (x)	Area $F(x)$
−3·2	0·0007	−1·1	0·1357	1·0	0·8413
−3·1	0·0010	−1·0	0·1587	1·1	0·8643
−3·0	0·0013	−0·9	0·1841	1·2	0·8849
−2·9	0·0019	−0·8	0·2119	1·3	0·9032
−2·8	0·0026	−0·7	0·2420	1·4	0·9192
−2·7	0·0035	−0·6	0·2743	1·5	0·9332
−2·6	0·0047	−0·5	0·3085	1·6	0·9452
−2·5	0·0062	−0·4	0·3446	1·7	0·9554
−2·4	0·0082	−0·3	0·3821	1·8	0·9641
−2·3	0·0107	−0·2	0·4207	1·9	0·9713
−2·2	0·0139	−0·1	0·4602	2·0	0·9772
−2·1	0·0179	0	0·5000	2·1	0·9821
−2·0	0·0228	0·1	0·5398	2·2	0·9861
−1·9	0·0287	0·2	0·5793	2·3	0·9893
−1·8	0·0359	0·3	0·6179	2·4	0·9918
−1·7	0·0446	0·4	0·6554	2·5	0·9938
−1·6	0·0548	0·5	0·6915	2·6	0·9953
−1·5	0·0668	0·6	0·7257	2·7	0·9965
−1·4	0·0808	0·7	0·7580	2·8	0·9974
−1·3	0·0968	0·8	0·7881	2·9	0·9981
−1·2	0·1151	0·9	0·8159	3·0	0·9987

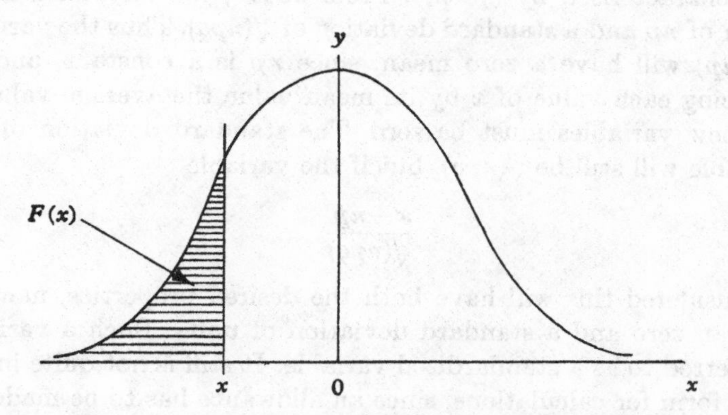

Fig. 9.4. Normal distribution, mean zero, standard deviation unity, area tabulated

9.8 Since table 9.4 only gives values appropriate to a mean of zero and a standard deviation of unity it cannot be used directly to calculate binomial probabilities. The mean and standard deviation of a binomial distribution are first required. It is found for any binomial distribution that

Mean number of successes is np,
Standard deviation of the number of successes is $\sqrt{(npq)}$.

These values can be verified for a particular case. Consider, for example the expansion

$$(q+p)^3,$$

with individual probabilities q^3, $3q^2p$, $3qp^2$, p^3. The mean number of successes will be

$$0\times q^2+1\times 3q^2p+2\times 3qp^2+3\times p^3 = 3p(q^2+2pq+p^2) = 3p,$$

since $$q^2+2qp+p^2 = (q+p)^2 = 1.$$

This agrees with the formula np when n is put equal to 3. The variance of the number of successes, applying formula (7.1), will be

$$(0-3p)^2q^3+(1-3p)^2\,3q^2p+(2-3p)^2\,3qp^2+(3-3p)^2p^3,$$

and this simplifies to $3pq$ remembering that $p+q = 1$. Hence the standard deviation is equal to $\sqrt{(3pq)}$ and agrees with the formula above taking n equal to 3.

Suppose now that x is the variable having the binomial distribution characterised by $(q+p)^n$. From above, the variable x has a mean of np and a standard deviation of $\sqrt{(npq)}$. Thus the variable $(x-np)$ will have a zero mean, since np is a constant, and by reducing each value of x by its mean value the average value of the new variables must be zero. The standard deviation of the variable will still be $\sqrt{(npq)}$, but if the variable

$$\frac{x-np}{\sqrt{(npq)}}$$

is considered this will have both the desired properties, namely, a mean zero and a standard deviation of unity. Such a variable is referred to as a standardised variable. It still is not quite in the ideal form for calculations, since an allowance has to be made for the fact that x only takes integral values, whereas the normal distribution is continuous and can take any value. By making use of the normal curve tables, a smooth continuous curve is being visualised so that some form of correction has to be made.

If the probability of just x successes is required, fig. 9.5 shows that a reasonable approximation is to take the area under the normal curve from $x-\frac{1}{2}$ to $x+\frac{1}{2}$ and assume that this is equal to the shaded area. To find the sum of a series of probabilities corresponding to x, $x+1$, $x+2$, ..., it is only necessary to find the area under the normal curve not from x to infinity but from $x-\frac{1}{2}$ to infinity. This correction of taking the area from a point $\frac{1}{2}$ unit nearer the centre than x is called a *continuity correction*.

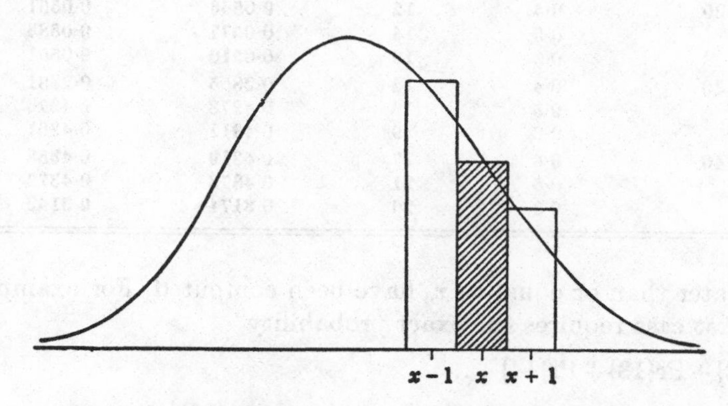

Fig. 9.5. Approximation to binomial with normal distribution

There are two cases to consider. For the probability of x or fewer successes, the function to be used is

$$\frac{x-np+\frac{1}{2}}{\sqrt{(npq)}},$$

whilst if the probability of x or more successes is required, the function used is

$$\frac{x-np-\frac{1}{2}}{\sqrt{(npq)}},$$

and the probability in table 9.4 is subtracted from one to give the right-hand tail. Clearly this procedure must result in an approximation to the true probabilities, as the normal curve has abscissae stretching theoretically from minus infinity to plus infinity, whereas the binomial distribution only goes from 0 to n inclusive. However, it is a very good approximation and is excellent for large values of n. The figures given in table 9.5 illustrate this. Here nine particular situations have been taken, and for each the exact probabilities, together with the approximate probabilities, that x

Table 9.5. *Comparison of binomial probabilities*

			$\Pr\{x \geqslant x_0\}$	
n	p	x_0	Exact binomial probability	Approximate normal probability
20	0·4	12	0·0566	0·0551
	0·5	14	0·0577	0·0588
	0·6	16	0·0510	0·0551
30	0·4	14	0·2855	0·2881
	0·5	16	0·4278	0·4276
	0·6	19	0·4311	0·4261
40	0·4	17	0·4319	0·4358
	0·5	21	0·4373	0·4372
	0·6	26	0·3174	0·3142

is greater than or equal to x_0 have been computed. For example, the first case requires the exact probability

$$\Pr\{12\}+\Pr\{13\}+\Pr\{14\}+\ldots$$

$$= \binom{20}{12}(0{\cdot}4)^{12}(0{\cdot}6)^{8}+\binom{20}{13}(0{\cdot}4)^{13}(0{\cdot}6)^{7}+\binom{20}{14}(0{\cdot}4)^{14}(0{\cdot}6)^{6}+\ldots$$

$$= 0{\cdot}0355+0{\cdot}0146+0{\cdot}0049+0{\cdot}0013+0{\cdot}0003+\ldots$$

$$= 0{\cdot}0566.$$

For the approximation, take

$$\frac{12-(20)(0{\cdot}4)-0{\cdot}5}{\sqrt{[(20)(0{\cdot}4)(0{\cdot}6)]}} = 1{\cdot}5975,$$

and using a more complete version of table 9.4 it is found that the area up to this abscissa is equal to 0·9449. Hence the area beyond the abscissa, which is the area required, will be $1-0{\cdot}9449$ or 0·0551. Comparing this with the exact value of 0·0566 we see that there is an error of $-0{\cdot}0015$ in the approximation. All the other values in the table can be obtained by similar processes. The approximation seems good for all the examples quoted, and if the number of trials, n, is large it improves. The approximation is valid provided p is not too close to 0 or 1 and the range of values of p for which satisfactory results are obtained is found to be roughly

$$\frac{9}{n+9} < p < \frac{n}{n+9}.$$

Thus if $n = 20$, the rule gives $0{\cdot}31 < p < 0{\cdot}69$, or if $n = 30$, the rule gives $0{\cdot}23 < p < 0{\cdot}77$, and in general the larger the value of n the wider the limits allowed for p.

9.9 The use of the above approximation will now be illustrated with an example. Remember that it is perfectly possible, although somewhat laborious, to calculate exactly the required probabilities from the binomial distribution.

Example 9.7 Over a period of time a large number of mice have been given an infection which is resistant to penicillin. A proportion 0·42 of the mice subsequently die from the infection within a week. A new type of antibiotic has been discovered and it is desired to see whether this antibiotic will reduce the number of deaths from the infection. To investigate this a group of sixty-five mice are randomly selected and given both the infection and the antibiotic. Of this group twenty-two subsequently die within a week of being given the infection. On the basis of these figures has the antibiotic produced a reduction in the death-rate from the infection?

If the antibiotic has had *no* effect the probabilities of 0, 1, 2, ... deaths occurring amongst a group of sixty-five mice given the infection will be the successive terms of the binomial expansion

$$(0{\cdot}58+0{\cdot}42)^{65} \quad \text{where} \quad n = 65, p = 0{\cdot}42.$$

To find out whether the antibiotic is effective the lower end of the distribution has to be examined, since if there were any beneficial effect it will result in fewer deaths occurring. Hence the probability required is the probability that twenty-two or fewer of the mice would die if no reduction in the chance of death had taken place and this requires the sum

$$\Pr\{22 \text{ die}\}+\Pr\{21 \text{ die}\}+\Pr\{20 \text{ die}\}+\ldots+\Pr\{0 \text{ die}\}.$$

To calculate all these twenty-three terms individually would involve a great deal of arithmetic, but since the values of n and p satisfy the conditions laid down in section 9.8 the normal approximation can be used. The first quantity to be calculated is

$$\frac{22-(65)(0{\cdot}42)+0{\cdot}5}{\sqrt{[(65)(0{\cdot}58)(0{\cdot}42)]}} = -1{\cdot}21.$$

Referring to table 9.4 the area of the normal curve up to the ordinate at the abscissa of $-1\cdot21$ is equal to $0\cdot1131$. This value is correct to four decimal places. Thus a result such as has been obtained here, or a more extreme one, would occur in over 11 per cent of experiments even if p were unchanged from $0\cdot42$, and this does not therefore seem to be such an unlikely happening that a confident statement of the superiority of the antibiotic can be made. If it were still desired to pursue the effect of the antibiotic, further data would have to be collected.

EXERCISES

9.1 Four dice are thrown simultaneously and the number of sixes, x, noted. x can take the five values 0, 1, ..., 4. Find the probabilities that x takes each of these five values. What is the most likely value of x; that is, the value of x with the highest probability?

9.2 Six pennies are tossed simultaneously and the number of heads obtained is noted. If the procedure is repeated sixty times how many times would you expect to have just one head, two heads, three heads, ..., six heads amongst the six coins tossed?

9.3 A marksman on average scores a bull with 40 per cent of his shots without using an arm-rest. Given an arm-rest the marksman fires ten shots of which six are bulls. Would you say that this result shows that the use of a rest improves the marksman's shooting or not?

9.4 Routine tests of glass bottles consist of subjecting them mechanically to a heavy test blow and seeing if they break under the blow. At present the breakage rate is 25 %. To test whether a new basic material used for the bottles is stronger, a sample of fifteen bottles is taken and each is given the test blow. Of the fifteen bottles only one breaks. Is this evidence that some improvement has been obtained in the strength of the bottles or not?

9.5 According to a certain mortality table the probability that a man aged thirty dies within thirty years is $0\cdot247$. Five hundred men aged 30 are selected at random and of these 110 die within thirty years. Is this evidence in accordance with the mortality table, or does it appear that the mortality table overestimates the rate of mortality?

9.6 In a packet of flower seeds it has been found in the past that on average one-third of the seeds give red, and the remainder white, flowers. A row of seven seeds is planted. Calculate the probability that the row will contain

(*a*) no red flowers;
(*b*) just one red flower;
(*c*) no white flowers.

9.7 Radio valves are tested by subjecting them to a large electric shock. Each shock has an independent chance of 0·8 of destroying the valve. How many shocks must be given to a valve in order that the probability of the valve being destroyed is at least 0·99?

9.8 Assume that boys and girls are born in equal numbers. Calculate the proportions of families with four children that have 0, 1, 2, 3 or 4 boys. What is the most probable number of boys, that is, the number of boys that has the highest probability? Calculate the mean and standard deviation of the distribution that you have obtained.

9.9 On an average one telephone out of four in a city business area is busy between 11 a.m. and 12 noon. If nine randomly selected numbers are called between the two times mentioned, find the probability that

(*a*) all are free;
(*b*) one and only one is unavailable;
(*c*) two or more of the numbers are unavailable.

9.10 The probability of a person getting no aces when dealt a hand of thirteen cards from an ordinary pack of fifty-two cards is 0·30. What is the probability that a person plays six hands of bridge and

(*a*) never gets an ace in any hand;
(*b*) always gets at least one ace in every hand?

9.11 An ordinary six-sided die is tossed 240 times and on forty-eight of the tossings a six is obtained. Is this result compatible with the die being unbiased?

9.12 An ordinary penny is tossed twenty times and gives twelve heads in the twenty tossings. Would you doubt the unbiased nature of the penny?

9.13 An ordinary penny is tossed 400 times and gives 240 heads in the 400 tossings. Would you doubt the unbiased nature of the penny? (Note that there are twenty times as many tossings and twenty times as many heads obtained in this case as there were in exercise 9.12. Why do the two questions seem to give different answers?)

9.14 Two boys each toss a true penny five times. Calculate the probability that they get the same number of heads.

9.15 In a biochemical experiment twenty insects were put in each of 100 jars. After being subjected to a fumigant for 3 hr. the number alive in each jar was counted.

No. alive	0	1	2	3	4	5	6	7	8	9	Total
No. of jars	3	8	11	15	16	14	12	11	9	1	100

Investigate whether it could be considered that each insect has a common chance p of surviving the fumigant. Do this by calculating the mean of the observed distribution and equating it to the mean of the binomial distribution $(q+p)^n$. The binomial should give the probabilities of 0, 1, 2, ... alive, provided there is a constant p. The value of n is known to be 20 as there are twenty insects in each jar, and as the mean is np, p may be calculated. Finally calculate the expected number of jars in which there are 0, 1, 2, ... insects alive by multiplying each probability by 100, the number of jars observed. A comparison between the observed and expected series of numbers can now be made.

9.16 Three hundred and twenty rows of seeds of a certain vegetable are incubated. Each row has five seeds, and after a certain period the number of rows in which 0, 1, 2, ..., 5 seeds germinated was counted.

No. of seeds germinated	0	1	2	3	4	5	Total
No. of rows	81	122	88	21	6	2	320

By fitting the appropriate theoretical distribution discuss whether the data are consistent with the assumption that the chance of an individual seed germinating in all the rows is the same. (Hint: equate the observed mean with the mean of $(q+p)^5$.)

9.17 Sixty-six litters each of five mice were examined and the number of female mice counted.

No. of female mice	0	1	2	3	4	5	Total
No. of litters	2	15	21	21	6	1	66

Do you think the data are consistent with the assumption that the chance of a mouse being female is the same in all litters?

9.18 A person is to be tested to see whether he can differentiate between the taste of two brands of cigarettes. If he cannot differentiate, it is assumed that the probability is one-half that he will identify a cigarette correctly. Under which of the following two procedures is there less chance that he will make all correct identifications when he actually cannot differentiate between the two tastes?

(*a*) The subject is given four pairs, each containing both brands of cigarettes (this is known to the subject); he must identify for each pair which cigarette represents each brand.

(*b*) The subject is given eight cigarettes and is told that the first four are of one brand and the last four of the other brand.

How do you explain the difference in results despite the fact that eight cigarettes are tested in each case?

10

FURTHER PROBABILITY CONCEPTS

10.1 In this chapter some of the probability theory developed in the last two chapters will be taken a stage further. The normal distribution was shown to be a limiting form of the binomial distribution; in this chapter another limiting form of the binomial, namely the Poisson distribution, will be discussed. Next a generalisation of the binomial distribution to the multinomial distribution, where more than two possible categories are involved, will be discussed. Thirdly the so-called Bayes' theorem in probability will be defined and illustrated, and finally the concept of statistical expectation is introduced.

10.2 The Poisson distribution arises as a limiting form of the binomial distribution $(q+p)^n$ in the situation when n tends to infinity and p tends to zero in such a manner that np remains constant and finite. (The ensuing proof is not essential to an understanding of the illustrations following it.)

It will be recalled that the probability of precisely k successes in n independent trials, at each of which the probability of a success is equal to p, is

$$P_k = \binom{n}{k} p^k(1-p)^{n-k}.$$

the expression may be written as

$$P_k = \frac{n!}{k!\,(n-k)!}\,p^k(1-p)^n/(1-p)^k.$$

Letting n tend to infinity and p tend to zero in such a way that the mean np remains fixed at some value m (so that $p = m/n$) and the variance npq $(= np(1-p))$ also tends to m gives:

$$\begin{aligned}
\lim_{n\to\infty} P_k &= \lim_{n\to\infty} \frac{n!}{k!\,(n-k)!}\,\frac{m^k}{n^k}\,\frac{(1-m/n)^n}{(1-m/n)^k} \\
&= \lim_{n\to\infty} \frac{n(n-1)\ldots(n-k+1)}{k!}\,\frac{m^k}{n^k}\,\frac{(1-m/n)^n}{(1-m/n)^k} \\
&= \frac{m^k}{k!}\lim_{n\to\infty}\frac{n(n-1)\ldots(n-k+1)}{n^k(1-m/n)^k}\,(1-m/n)^n \\
&= \frac{m^k}{k!}\lim_{n\to\infty}(1-m/n)^n = \frac{m^k}{k!}\,e^{-m},
\end{aligned}$$

where e is the base of natural logarithms (having the value of 2·71828). Denoting the limit for convenience by $p(k)$ gives:

$$p(k) = \frac{m^k}{k!} e^{-m}$$

which is the probability distribution commonly known as the *Poisson* distribution. It can be used to approximate to P_k for large n and small p as follows:

$$P_k \doteqdot \frac{(np)^k e^{-np}}{k!}.$$

Note that:

$$\sum_{k=0}^{\infty} P_k \doteqdot \sum_{k=0}^{\infty} p(k) = e^{-m}\left(1+m+\frac{m^2}{2!}+\dots\right)$$
$$= e^{-m}.e^{m} = 1,$$

i.e. the sum of $p(k)$ over all the possible values of k is unity. The mean and variance of the Poisson distribution are both equal to m. Hence the standard deviation is $\sqrt{m}$. When fitting a Poisson distribution to observed data, it is usual to equate the mean of the observed distribution to m, the parameter of the corresponding Poisson distribution.

Example 10.1 Two per cent of the very large number of screws made by a machine are defective, the defectives occurring at random during production. If the screws are packaged 100 per box, what is the probability that a given box will contain exactly k defectives?

The probability that the box contains k defectives is given exactly by the binomial distribution

$$P_k = \binom{100}{k}\left(\frac{2}{100}\right)^k\left(1-\frac{2}{100}\right)^{100-k} \quad \text{for} \quad k = 0, 1, 2, \dots, 100.$$

Since $n = 100$, $p = 0{\cdot}02$ and $np = 2$, the Poisson approximation to P_k seems reasonable and is given by:

$$p(k) \doteqdot \frac{2^k e^{-2}}{k!}.$$

Values of this expression can be calculated easily with the aid of table 10.1 which gives specimen values of the exponential function.

Table 10.1. *Values of e^{-x}*

x	e^{-x}	x	e^{-x}	x	e^{-x}
0·00	1·00000	1·80	0·16530	4·20	0·01500
0·01	0·99005	2·00	0·13534	4·40	0·01228
0·05	0·95123	2·20	0·11080	4·60	0·01005
0·10	0·90484	2·40	0·09072	4·80	0·00823
0·20	0·81873	2·60	0·07427	5·00	0·00674
0·40	0·67032	2·80	0·06081	5·20	0·00552
0·60	0·54881	3·00	0·04979	5·40	0·00452
0·80	0·44933	3·20	0·04076	5·60	0·00370
1·00	0·36788	3·40	0·03337	5·80	0·00303
1·20	0·30119	3·60	0·02732	6·00	0·00248
1·40	0·24660	3·80	0·02237	6·20	0·00203
1·60	0·20190	4·00	0·01832	6·40	0·00166

A comparison between the numerical results obtained from evaluating both P_k and $p(k)$ for each value of k in this particular example is given in table 10.2. The agreement between the exact binomial probability and the (approximate) Poisson probabilities is seen to be extremely good, the maximum error in an individual term being 0·0027.

Table 10.2. *Comparison of P_k and $p(k)$ for $n = 100$, $p = 0{\cdot}02$*

k	P_k	$p(k)$
0	0·1326	0·1353
1	0·2707	0·2707
2	0·2734	0·2707
3	0·1823	0·1804
4	0·0902	0·0902
5	0·0353	0·0361
6	0·0114	0·0120
7	0·0031	0·0034
8	0·0007	0·0009
9	0·0002	0·0002

Example 10.2 It is known that 0·006 per cent of the insured males in a particular country die from a certain kind of accident each year. What is the probability that an insurance company must pay off on more than 3 of 10,000 insured risks against such accidents in a given year?

Here $p = 0{\cdot}00006$ and $n = 10{,}000$, so that $m = np = 0{\cdot}6$. Hence the required probability is:

$$\sum_{k=4}^{k=\infty} p(k) = e^{-m}\left(\frac{m^4}{4!}+\frac{m^5}{5!}+\frac{m^6}{6!}+\dots\right)$$

$$= 0{\cdot}54881(0{\cdot}00540+0{\cdot}00065+0{\cdot}00006+0{\cdot}00001+\dots)$$

$$= 0{\cdot}54881 \times 0{\cdot}00612$$

$$= 0{\cdot}00336.$$

The value of e^{-m} has again been taken from table 10.1

Table 10.3. *Distribution of flying-bomb landings*

No. of flying-bombs per square	Observed no. of squares	Expected no. of squares
0	229	226·7
1	211	211·4
2	93	98·5
3	35	30·6
4	7	7·1
5 or over	1	1·6
Totals	576	575·9

Example 10.3 An unusual illustration of the Poisson distribution was given by R. D. Clarke in the *Journal of the Institute of Actuaries*, vol. 72, 1946, p. 481. This related to an investigation made to test whether the points of impact of flying-bombs in London tended to be grouped in clusters. An area of 144 sq. km. in South London was divided into 576 squares each of $\frac{1}{4}$ sq. km. and a count made of the numbers of squares within which 0, 1, 2, ... flying bombs fell. These figures are given in the third column of table 10.3. As 537 bombs in all fell within the total area, it was assumed that the value of m for the appropriate distribution was $\frac{537}{576}$, i.e. the average number of bombs per square. The expected number of squares within which k bombs fell were then calculated from the expression

$$576\,\frac{m^k}{k!}\,e^{-k}$$

where $m = \frac{537}{576}$. These numbers are shown in the final column of table 10.3 and conform very well with the actual numbers observed.

This tends to support the hypothesis that the bombs fell at random within the whole area concerned, since the numbers falling per square are consistent with a mathematical model whereby each bomb had the same chance of falling in any particular square.

10.3 Suppose that a trial can result in one and only one of r mutually exclusive events $E_1, E_2, \ldots, E_r$, with probabilities $p_1, p_2, \ldots, p_r$ respectively, where

$$\sum_{i=1}^{r} p_i = 1.$$

If n independent trials are made then, by an argument similar to that used in chapter 9 to derive the binomial distribution, the probability of obtaining precisely k_1 E_1's, k_2 E_2's, etc., can be shown to be given by:

$$p(k_1, k_2, \ldots, k_r) = \frac{n!}{k_1!\, k_2! \ldots k_r!} p_1^{k_1} p_2^{k_2} \ldots p_r^{k_r},$$

where $0 \leqslant k_i \leqslant n, i = 1, 2, \ldots, r$ and $\sum_{i=1}^{r} k_i = n$. This is the probability distribution known as the *multinomial distribution*. The name derives from the fact that the probabilities are the various terms in the expansion of $(p_1+p_2+\ldots+p_r)^n$. Notice that if $r = 2$, the distribution reduces, as is to be expected, to the binomial distribution and hence the multinomial distribution can be regarded as an extension of the binomial distribution.

Example 10.4 Ball-bearings of a particular type are being made, whose diameters should be 0·2500 inch. Because of the inherent variability in the manufacturing process, and because of consumer demands, the bearings are classified as undersize, oversize and acceptable if they measure less than 0·2495 inch, more than 0·2505 inch, and between 0·2495 and 0·2505 inch respectively.

Suppose that the production process for these bearings is such that 4 per cent of the bearings are undersize, 6 per cent are oversize and 90 per cent are acceptable. If 100 of these bearings are picked at random, the probability of getting k_1 undersize, k_2 oversize and k_3 acceptable bearings is given by:

$$p(k_1, k_2, k_3) = \frac{100!}{k_1!\, k_2!\, k_3!} (0{\cdot}04)^{k_1} (0{\cdot}06)^{k_2} (0{\cdot}90)^{k_3},$$

where $0 \leqslant k_1, k_2, k_3 \leqslant 100$ and $\sum_{i=1}^{3} k_i = 100$. Thus, if $k_1 = 5$, $k_2 = 6$ and $k_3 = 89$, then:

$$p(5, 6, 89) = \frac{100!}{5!\,6!\,89!}(0{\cdot}04)^5\,(0{\cdot}06)^6\,(0{\cdot}90)^{89}$$

$$= 0{\cdot}0256.$$

Compare this probability with the result if 2 per cent are undersize and 8 per cent are oversize, with 90 per cent still acceptable, when

$$p(5, 6, 89) = 0{\cdot}0037.$$

Example 10.5 Ten people each toss two coins. What is the probability that: (i) Three people throw two heads, three people throw two tails and four people throw one head and one tail? (ii) No one throws a head and a tail? For one individual the probability of two heads, one head and one tail, or two tails are, respectively, $\frac{1}{4}$, $\frac{1}{2}$, $\frac{1}{4}$. Call these three probabilities p_1, p_2, p_3. Then the required probability under (i) is:

$$\frac{10!}{3!\,4!\,3!}(\tfrac{1}{4})^3\,(\tfrac{1}{2})^4\,(\tfrac{1}{4})^3 = \tfrac{525}{8192} = 0{\cdot}0641.$$

There are several ways of obtaining the answer to (ii). It can be computed as the sum of eleven multinomial terms, the first being ten 'two tails' and no 'two heads', the next nine 'two tails' and one 'two heads', and so on. The probabilities are then:

$$\binom{10}{0}(\tfrac{1}{4})^{10} + \binom{10}{1}(\tfrac{1}{4})^{10} + \binom{10}{2}(\tfrac{1}{4})^{10} + \ldots + \binom{10}{10}(\tfrac{1}{4})^{10} = (\tfrac{1}{4})^{10}(1+1)^{10}$$

$$= (\tfrac{1}{4})^{10}\,2^{10} = (\tfrac{1}{2})^{10}.$$

Alternatively the problem is reduced to a binomial probability by having just the two categories: one head and one tail, or two similar. The two categories have equal probabilities and the required probability of none in the first category is $(\frac{1}{2})^{10}$ as before.

10.4 Note that both the binomial and multinomial distributions only apply to situations involving a series of independent trials. Thus if a bag contains m balls of which n are white, then the binomial distribution gives the probability of obtaining exactly t white balls and $(r-t)$ black balls in r random drawings as:

$$\binom{10}{t}\left(\frac{n}{m}\right)^t\left(1-\frac{n}{m}\right)^{r-t}$$

provided that each ball is replaced and the bag well shuffled before the next drawing takes place. If, however, the drawn balls are not replaced the appropriate probability could be computed as follows. There are $\binom{m}{r}$ ways of drawing r balls out of the bag of m. There are $\binom{n}{t}$ ways of selecting t white balls out of the n white balls, and $\binom{m-n}{r-t}$ ways of selecting $(r-t)$ black balls out of the $(m-n)$ black balls. Hence the probability that exactly t white balls and $(r-t)$ black balls are obtained in the r drawings is:

$$\binom{n}{t}\binom{m-n}{r-t}\Big/\binom{m}{r}.$$

It can be shown that when n is very large this probability tends to the binomial probability given earlier. This is to be expected since the probability of drawing a white ball under such circumstances remains almost constant at n/m. This distribution is referred to as the *hypergeometric* distribution.

10.5 If x is some variable which has a probability distribution, then x is referred to as a *random variable*. Thus if x is the number of pips showing on a die when tossed, then x is a random variable with the probability distribution:

$$p(x) = \tfrac{1}{6} \quad (x = 1, 2, \ldots, 6).$$

The mean, or *expectation*, of the random variable x is defined as:

$$E(x) = \sum_{i=1}^{k} x_i\, p(x_i),$$

where the random variable takes the k values $x_1, x_2, \ldots, x_k$. Thus for the Poisson distribution discussed earlier

$$p(x) = \frac{e^{-m}m^x}{x!} \quad (x = 0, 1, 2, \ldots).$$

Hence
$$\begin{aligned} E(x) &= \sum_{x=0}^{\infty} xp(x) \\ &= \sum_{x=0}^{\infty} x\frac{e^{-m}m^x}{x!} = me^{-m}\sum_{x=1}^{\infty}\frac{m^{x-1}}{(x-1)!} \\ &= m. \end{aligned}$$

Note that the term expectation must not be confused with the modal value, i.e. it is not necessarily the most likely or expected value in that sense. It is equated with the mean value.

Example 10.6 With a true and unbiased roulette wheel there are 37 slots, 18 marked red, 18 marked black, and one green. A man backs a colour (red or black) with £1; he receives £1 plus his stake if he wins: otherwise he loses his stake. What is his expectation per play?

The probability of winning is $\frac{18}{37}$, and of losing $\frac{19}{37}$. Hence his expectation is

$$2 \times \tfrac{18}{37} + 0 \times \tfrac{19}{37} - 1 = -\tfrac{1}{37}.$$

This does not, of course, imply that the man loses £$\frac{1}{37}$ on each throw—indeed he must either make £1 or lose £1. What it does mean, however, is that over a large number of throws, say N, the loss will approximate to £$N/37$, the likely variation observed about this expectation being smaller proportionally the larger the size of N.

Example 10.7 Packets of tea contain one of a set of 50 different types of card depicting flags of various nations. A collector has already got 30 of the 50 different possible cards. What is the expected number of packets of tea he will have to purchase before his 31st card in the set is obtained? (Assume that there are effectively an infinite but equal number of each card available and that the cards are inserted at random in the packets.)

The probability, at each purchase of a packet of tea, that a new card will be found is $\frac{20}{50}$ as there are 20 new cards left in the set. If r packets have to be bought before a new card is found, this implies that the first $(r-1)$ purchases give old cards and the rth purchase gives a new card. The probability of this occurring is:

$$\left(\frac{30}{50}\right)^{r-1}\left(\frac{20}{50}\right).$$

Hence the expected value of r is:

$$E(r) = \sum_{r=1}^{\infty} r\left(\frac{30}{50}\right)^{r-1}\left(\frac{20}{50}\right)$$

$$= \frac{20}{50}\sum_{r=1}^{\infty} r\left(\frac{30}{50}\right)^{r-1}$$

$$= \frac{20}{50}\left[1+2\left(\frac{30}{50}\right)+3\left(\frac{30}{50}\right)^2+4\left(\frac{30}{50}\right)^3+\dots\right].$$

Let the portion inside the square brackets be denoted by S. Then

$$\left(\frac{30}{50}\right) S = \frac{30}{50}+2\left(\frac{30}{50}\right)^2+3\left(\frac{30}{50}\right)^3+\dots$$

and

$$S-\left(\frac{30}{50}\right) S = 1+\frac{30}{50}+\left(\frac{30}{50}\right)^2+\left(\frac{30}{50}\right)^3+\dots$$

$$= \frac{1}{1-\frac{30}{50}} = \frac{50}{20} \quad \text{(as the sum of an infinite geometric series).}$$

Hence $$S = \frac{50}{20}\Big/\left(1-\frac{30}{50}\right) = \frac{50}{20}.\frac{50}{20}.$$

This gives $$E(r) = \frac{20}{50}S = \frac{50}{20} = 2{\cdot}5.$$

10.6 So far probability theory has been used to derive some basic rules, whereby the probability of the happening of one event can be derived from the probabilities of other events. The assumption has been that, from what are termed *prior* probabilities, the probabilities of certain happenings can be computed. Assume now that some experiment has been performed, or sample drawn, and the results of this experiment are known. What are now the probabilities, termed *posterior*, corresponding to the original prior probabilities? An illustration will help to make the position clear.

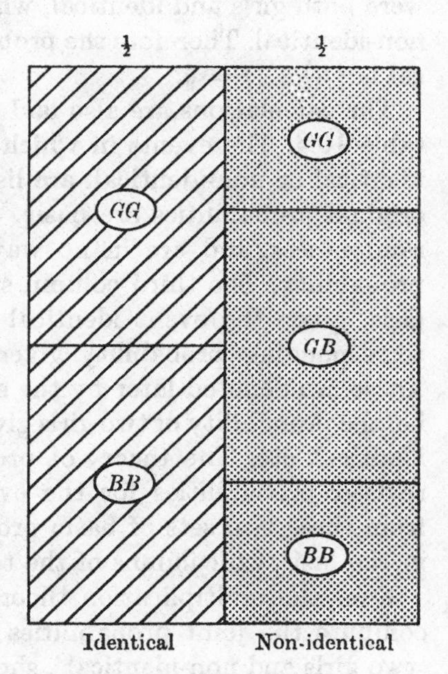

Fig. 10.1. Composition of twin population, G = Girl, B = Boy

Suppose that twins are of two types, identical (I) and non-identical (N) and that in the population as a whole there are 50 per cent of each type. Secondly, suppose that identical twins are always of the same sex, with a 50 per cent chance of each sex occurring, whilst for non-identical twins the sex of each child has an independent 50:50 chance of being male or female. The mix that would result from these suppositions is shown pictorially in figure 10.1.

Suppose now that a pair of twins has been selected at random from the population and both are girls. What is the posterior probability that the pair of twins is an identical pair? It is unlikely that you would still assign equal probabilities to the twins being identical or non-identical, since more identical sets of twins have two girls than do non-identical sets of twins. The argument can then proceed as follows.

Originally every twin in the population was assigned equal probability. The subsequent information that the pair of twins were both girls obviously leads to a change in this assessment so that zero probability is assigned to all boys, and hence raises the total probability assigned to all pairs of girl twins to 1. There is, however, no reason to assign a higher probability to any one pair of girl twins than to any other and hence all pairs of girl twins are given equal probability. Of all twins in the population, $\frac{1}{2}\times\frac{1}{2} = \frac{1}{4}$ were both girls and identical, whilst $\frac{1}{2}\times\frac{1}{4} = \frac{1}{8}$ were both girls and non-identical. Therefore the probability that the twins were identical is $\frac{1}{4}/(\frac{1}{4}+\frac{1}{8}) = \frac{2}{3}$.

The calculations are also laid out in the tabular form shown in table 10.4. The events in which interest is concentrated, namely identical or non-identical, are listed in the first column and their original probabilities are basic, depending upon the definition of randomness, and are in no way dependent upon the theory of probability. The third column shows first the probability of two girls, given the event identical twins, and secondly that of the corresponding probability given the event non-identical twins. These are denoted later by the symbols $\Pr\{GG|I\}$ and $\Pr\{GG|N\}$, i.e. the probability of two girls given that the twins are known to be identical, etc. The theory of probability is then used to compute revised probabilities for the events identical and non-identical from these two sets of basic probabilities, the work being shown in the last two columns of the table.

First the multiplication theorem (II of section 8.4) is used to compute the joint probabilities of 'two girls and identical' and 'two girls and non-identical', shown in the penultimate column as assessed before it was known that the twins were both girls. The addition theorem (I of section 8.3) is next used to compute the

Table 10.4. *Calculation of revised probabilities*

	Original probabilities		Revised probabilities	
Event of interest (1)	Probability of event in (1) (2)	Probability of two girls given event in (1) (3)	Joint probability of two girls and event in (1) (4) = (2) × (3)	Probability of event in (1) given two girls (5) = (4) ÷ Σ(4)
Identical	$\frac{1}{2}$	$\frac{1}{2}$	$\frac{1}{4}$	$\frac{2}{3}$
Non-identical	$\frac{1}{2}$	$\frac{1}{4}$	$\frac{1}{8}$	$\frac{1}{3}$
Totals	1	—	$\frac{3}{8}$	1

marginal probability of two girls as assessed before it was known that the twins were both girls; this is the $\frac{3}{8}$ total of the joint probabilities. Finally the last column uses the definition of conditional probability to compute the revised probabilities. For this theorem II needs to be extended to cover two events that may or may not be independent. Written symbolically the theorem now reads:

$$\Pr\{AB\} = \Pr\{A\}\Pr\{B|A\}$$

$$\text{or} \quad = \Pr\{B\}\Pr\{A|B\}.$$

In words, $\Pr\{B|A\}$ may be said to be the conditional probability of the event B, given the event A; or the probability assigned to B when it is known that A has occurred.

10.7 Returning to the previous illustration, the logic in table 10.4 can now be expressed more compactly by the use of algebraic notation. First, from the now extended multiplication theorem:

$$\Pr\{GG,I\} = \Pr\{I\}\Pr\{GG|I\}$$

and

$$\Pr\{GG,N\} = \Pr\{N\}\Pr\{GG|N\}.$$

Next the addition theorem shows that:

$$\Pr\{GG\} = \Pr\{I\}\Pr\{GG|I\} + \Pr\{N\}\Pr\{GG|N\}$$

and finally the extended multiplication theorem gives:

$$\Pr\{I|GG\} = \frac{\Pr\{I, GG\}}{\Pr\{GG\}} = \frac{\Pr\{I\}\Pr\{GG|I\}}{\Pr\{I\}\Pr\{GG|I\} + \Pr\{N\}\Pr\{GG|N\}}.$$

This expression is known as *Bayes' theorem*, after the clergyman T. Bayes (1702–61) who discovered it, but it is really little more than an extension of the definition of conditional probability. In this instance only two events were concerned, but it can be straightforwardly generalised as follows:

If E_i $(i = 1, 2, \ldots, r)$ are r mutually exclusive and only possible events such that an event F can occur only if one of these r events happens, then the probability that E_j happens when F is known to have happened is

$$\Pr\{E_j|F\} = (\Pr\{E_j\} \Pr\{F|E_j\}) \Big/ \sum_{i=1}^{r} \Pr\{E_i\} \Pr\{F|E_i\}.$$

Two further examples of the application of this theorem will now be given.

Example 10.8 A dealer is buying a batch of machine parts from a manufacturer. He knows that the parts are produced by either machine I or machine II, but he cannot tell which. From past experience he knows that if machine I produced the parts 15 per cent will be defective, whilst if machine II produced them 25 per cent will be defective. The manufacturer reports that 60 per cent of batches come from machine I and 40 per cent from machine II. If the dealer selects a random sample of three parts from the assumed large batch and finds none defective, what is the (posterior) probability that the batch was produced by machine I?

Here let E_1 and E_2 represent the two machines and F the event that, with a random sample of three parts, no defectives are found. From the data the following probabilities can be written down:

$$\Pr\{E_1\} = 0{\cdot}6, \qquad \Pr\{E_2\} = 0{\cdot}4,$$
$$\Pr\{F|E_1\} = (1-0{\cdot}15)^3 = 0{\cdot}85^3,$$
$$\Pr\{F|E_2\} = (1-0{\cdot}25)^3 = 0{\cdot}75^3.$$

Hence

$$\Pr\{E_1|F\} = \frac{\Pr\{E_1\} \Pr\{F|E_1\}}{\Pr\{E_1\} \Pr\{F|E_1\} + \Pr\{E_2\} \Pr\{F|E_2\}}$$
$$= \frac{0{\cdot}6 \times 0{\cdot}85^3}{0{\cdot}6 \times 0{\cdot}85^3 + 0{\cdot}4 \times 0{\cdot}75^3}$$
$$= 0{\cdot}686.$$

Note that the bringing in of the sample information has lifted the initial *prior* probability of 0·6 to a figure of 0·686.

Example 10.9 A book club sells paperback reprints of scientific books by mail order to its members only. For a new book,

ABC, the book club will make a profit provided that it sells copies to more than 4,000 of its 100,000 members. Past experience has shown that (in a simplified form) the sales of this kind of book are likely to fall into one of three categories with the stated probabilities:

Event	Proportion of members purchasing	Prior probability
E_1	0·01	0·04
E_2	0·03	0·30
E_3	0·05	0·66

The publisher, not satisfied with the *prior* chance of 0·66 of the financial success that he would get if event E_3 turns out to be true, decides to send out an advance card to 100 of his membership, randomly selected, and ask them whether they will purchase the book. (Any orders received can be met, even if the publisher does not go ahead with the paperback edition, by supplying hard cover copies.) Of the 100 cards sent out, 6 result in an order for ABC. What is the posterior probability of E_3 (the only financially rewarding event)?

Here $\Pr\{E_1\} = 0{\cdot}04, \quad \Pr\{E_2\} = 0{\cdot}30, \quad \Pr\{E_3\} = 0{\cdot}66$

and
$$\Pr\{F|E_1\} = \binom{100}{6}(0{\cdot}01)^6\,(0{\cdot}99)^{94},$$
$$\Pr\{F|E_2\} = \binom{100}{6}(0{\cdot}03)^6\,(0{\cdot}97)^{94},$$
$$\Pr\{F|E_3\} = \binom{100}{6}(0{\cdot}05)^6\,(0{\cdot}95)^{94}.$$

Hence
$$\Pr\{E_3|F\} = \frac{\Pr\{E_3\}\Pr\{F|E_3\}}{\Pr\{E_1\}\Pr\{F|E_1\}+\Pr\{E_2\}\Pr\{F|E_2\}+\Pr\{E_3\}\Pr\{F|E_3\}}$$
$$= \frac{0{\cdot}66\binom{100}{6}(0{\cdot}05)^6\,(0{\cdot}95)^{94}}{0{\cdot}04\binom{100}{6}(0{\cdot}01)^6\,(0{\cdot}99)^{94}+0{\cdot}30\binom{100}{6}(0{\cdot}03)^6\,(0{\cdot}97)^{94}+0{\cdot}66\binom{100}{6}(0{\cdot}05)^6\,(0{\cdot}95)^{94}}$$
$$= 0{\cdot}868.$$

which is considerably higher than the original *prior* probability of 0·66 and might well incline the publisher more firmly in favour of the paperback venture.

EXERCISES

10.1 On average, 1 house in 2,000 in a particular district has a fire during a year. If 4,000 houses are in that district, what is the probability that exactly 5 houses will have a fire during the year?

10.2 A book of 500 pages contains 500 misprints scattered at random. Estimate the chance that a selected page contains at least three misprints.

10.3 Colour-blindness appears in 1 per cent of the people in a certain large population. How large must a random sample be if the probability of its containing at least one colour-blind person is to be 0·95 or more?

10.4 A small car-hire firm has two cars which it rents out by the day. Suppose that the number of demands for a car on each day is distributed as a Poisson distribution with mean 1·5.

(*a*) On what proportion of days is neither car required?

(*b*) On what proportion of days is the demand in excess of the company's capacity?

10.5 A total of 1,000 students participated in a spelling contest. The table below gives N_k, the number of students with k mis-spelled words out of the 100 that were presented for spelling to each student. Calculate the expected frequencies corresponding to the N_k if it is assumed that the probability of a mis-spelling is constant for each student and each word presented.

k	0	1	2	3	4	5	6	7	8	9
N_k	50	145	225	230	165	105	50	20	10	0

10.6 In a box of beads, 40 are red, 35 are green and 25 are black. Five beads are randomly selected, the bead being returned each time before the next is drawn. What is the probability that the five beads contain just three red, one green and one black?

10.7 Human blood groups have been classified in four exhaustive, mutually exclusive categories, O, A, B, AB. In a very large population, the respective probabilities of these four types are 0·46, 0·40, 0·11 and 0·03. Given that the population is large enough to be regarded as infinite, find the probabilities that a random sample of five individuals will contain:

(*a*) two cases of type O and one case of each of the others,

(*b*) three cases of type O and two of type A, or

(*c*) no case of type AB.

10.8 Competitors A, B and C control respectively 50 per cent, 30 per cent and 20 per cent of the market for a certain commodity. In a random sample of four buyers:

(*a*) what trade division amongst the three competitors is the most likely (i.e. has the highest probability)?

(*b*) what division is the next most likely (i.e. has the second highest probability)?

10.9 In the experience of a certain insurance company, customers who have sufficient funds in their bank post-date a cheque by mistake one in 1,000 times, whilst customers who write cheques on insufficient funds invariably post-date them. The latter group constitutes 1 per cent of the total. A company receives a post-dated cheque from a policy-holder. What is the probability that such a customer has insufficient funds?

10.10 Stores A, B and C under single ownership have 50, 75 and 100 employees and respectively 50 per cent, 60 per cent and 70 per cent of these are women. Resignations are equally likely to occur amongst all employees, regardless of sex. One employee resigns and this is a woman. What is the probability that she works in store C?

10.11 There are three boxes, each with two drawers. The drawers contain coins as follows: in the first box each drawer contains a gold coin; in the second box each drawer contains a silver coin; and in the last box a gold and a silver coin are contained in its two drawers, respectively. If equal probability selection methods are used to select a box, then a drawer, what is the probability that if a gold coin is selected it was obtained from box 3?

10.12 An insurance company writes a policy for an amount of money S, which is payable if an event E occurs. If the probability of event E occurring is p, what premium should the company charge the customer in order that the expected profit to the company is 5 per cent of S?

10.13 Show that, for a binomial variable k, where

$$P_k = \binom{n}{k} p^k(1-p)^{n-k}$$

the value of $E(k) = np$.

10.14 A soap company distributes blank entry forms to a lottery requiring nothing more than filling in of one's name and the posting of the form. The prize schedule is:

1st prize	£5,000	Next 10 prizes	£100 each
2nd prize	£2,000	Next 50 prizes	£50 each
3rd prize	£1,000	Next 100 prizes	£20 each
Next 5 prizes	£200 each	Next 1,000 prizes	£5 each

Assume that 10 million entry forms are returned. Let the random variable x be your gain from participation with a single entry.

(*a*) What is $E(x)$?

(*b*) If it costs you $1\frac{1}{2}$p to post the entry form, does it make sense to participate?

11

TESTS OF SIGNIFICANCE

11.1 In chapter 9 attention was focused on the reduction of the data in a problem into the form of probability, so that judgment can be made as to the likelihood of occurrence of observed results. The normal curve, or normal distribution, was introduced as the limiting form of the binomial distribution when n increases. This limiting form has a mathematical background and it can be derived in a theoretical manner. However, the normal distribution also turns up in many other ways, and it is this feature that gives it a central position in statistical theory today. Many of the distributions of height or weight or breadth of animals and plants, for example, or of the dimensions of manufactured articles, are found to be very close to the normal distribution in shape. This does not mean that the measured values follow a normal distribution exactly, but that they do so sufficiently closely to make the normal distribution a reasonable starting-point for statistical calculations and deductions. Other advantages are that the normal distribution is easily handled mathematically, has certain well-defined properties, and has been extensively tabulated. Referring to the diagram of the normal distribution in fig. 9.4 and labelling the co-ordinates (x, y) the equation of the curve is

$$y = \frac{1}{\sqrt{(2\pi)}s} e^{-\frac{1}{2}\left(\frac{x-\bar{x}}{s}\right)^2}, \tag{11.1}$$

where $\bar{x}$ and s are the mean and standard deviation of the distribution, and e is the base of natural logarithms (the constant 2·71828). Thus table 9.4 is calculated on the basis that $\bar{x}$ is zero and s is equal to one. To find corresponding normal distributions with different values of $\bar{x}$ and s is a fairly straightforward matter. If s is kept constant the spread of the distribution remains the same but the location of it will change. This follows because if a set of observations are all increased by a constant amount their mean value is increased by that amount but the standard deviation is unaltered. From (11.1) the value of y for a constant value of s depends only on the quantity $(x-\bar{x})$. This quantity is unchanged if all the values of x are altered by a constant amount, so that $\bar{x}$ alters the position of the distribution but not its shape or scale. Next consider $\bar{x}$ to be fixed at zero and see the effect of altering s. Straightforward calculations with (11.1) can be made taking $\bar{x}$ to be zero and s to have the values 1, 2 and 3. The results are shown in fig. 11.1 where the area under the three curves is still unity and

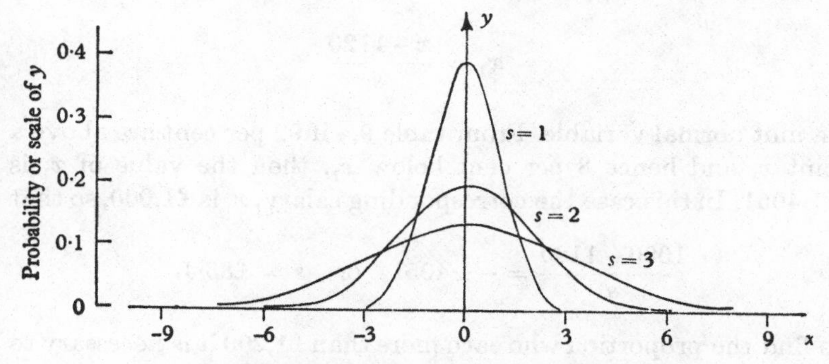

Fig. 11.1. Normal distributions with mean zero

each has the same mean value, namely zero. However, the normal distribution with the largest standard deviation is more spread out than the others, and in fact the standard deviation is essentially a measure of the spread of the distribution. Although the ordinates never quite become zero at either extremity the bulk of the distribution is seen to be contained

between -3 and $+3$ when $s = 1$,

between -6 and $+6$ when $s = 2$,

between -9 and $+9$ when $s = 3$,

and in general it is found that the distribution is approximately contained between $-3s$ and $+3s$.

11.2 The properties of the normal distribution can sometimes be used the other way round in order to deduce the mean or standard deviation given some other facts about the distribution. An example will demonstrate the technique.

Example 11.1 The distribution of salaries of the members of a certain firm is known to be in the form of a normal distribution with a mean of £1,120 per annum. Members of the staff earning a salary of £1,000 per annum or more must join the superannuation fund and it is found that 92 per cent of the members do in fact belong to the fund. What proportion of members earn more than £1,200 per annum?

Suppose that the unknown standard deviation is s and that x is a member selected at random. Then it follows that

$$x_1 = \frac{x-1120}{s}$$

is a unit normal variable. From table 9.4 if 92 per cent are above a point x_1 and hence 8 per cent below x_1, then the value of x_1 is $-1{\cdot}4051$. In this case the corresponding salary, x, is £1,000, so that

$$\frac{1000-1120}{s} = -1{\cdot}4051 \quad \text{or} \quad s = \pounds 85{\cdot}4.$$

To find the proportion who earn more than £1,200 it is necessary to convert £1,200 into a unit normal variable. This gives

$$\frac{1200-1120}{85{\cdot}4} = +0{\cdot}9368,$$

and from table 9.4 the area to the right of the ordinate at $+0{\cdot}9368$ is $0{\cdot}1745$. Hence some $17\frac{1}{2}$ per cent of the members will have salaries in excess of £1,200 per annum.

11.3 In chapter 8 samples of various sizes were drawn from a population and it was observed that if the standard deviation of the original observations was s, that of the means of the samples of size n was equal to $s/\sqrt{n}$. It was also noticeable that as the size of the sample increased the means of the samples tended to be normally distributed even though the original variables were not themselves normally distributed. It is a mathematical property that if the original variables are normally distributed the sample means are similarly distributed but with altered constants. Thus

Individuals: normally distributed mean $\bar{x}$, standard deviation s.

Means of samples of n: normally distributed mean $\bar{x}$, standard deviation $s/\sqrt{n}$.

This property is exact if the variables come from a normal distribution but otherwise only approximate, although the approximation improves very rapidly as n increases. When examining the means of samples, it is legitimate to assume them to be approximately normal in their distribution even if it is known that the original observations were not so distributed. For convenience of description it is useful to distinguish between constants that belong to the population of individuals being sampled and constants calculated from the sample values obtained by sampling the population. It is customary to use Greek letters for the former and italic letters for the latter giving: ξ (xi) for the population mean, $\bar{x}$ for the sample mean; σ (sigma) for the population standard deviation, s for the sample standard deviation.

Suppose then that a normal population is specified by the values ξ and σ. Then if random samples of size n are drawn, the mean, $\bar{x}$, of such samples will itself have a distribution that is normal. The constants of the distribution will now be ξ and $\sigma/\sqrt{n}$. That is, it will still be located at the same place but will be much more pinched in appearance.

In the examples of chapter 9, such as 9.6, the questions asked were of the form 'is the probability of an event p, or has it increased?' Now that the interest concerns a measurable characteristic, and not just presence or absence form of characteristic, a typical question would be 'does this set of observations come from a population with mean ξ_0 or does it come from a population whose mean is ξ_1 where ξ_1 is greater than ξ_0?' As an example take $\xi_0 = 10$ and $\xi_1 = 12$ with σ equal to 1. In fig. 11.2 two distributions are given:

(i) normal distribution mean 10, standard deviation 1;

(ii) normal distribution mean 12, standard deviation 1.

These two distributions overlap a great deal. Thus if one observation is drawn from the population and its value is 10·4 it is impossible to say whether it comes from the distribution specified by (i) or the distribution specified by (ii) since it is quite a likely occurrence in either case. But suppose that instead of drawing just one observation and measuring it a sample of sixteen observations is drawn, each observation measured and the mean of the sixteen observations found to be 11.4. The distributions of the

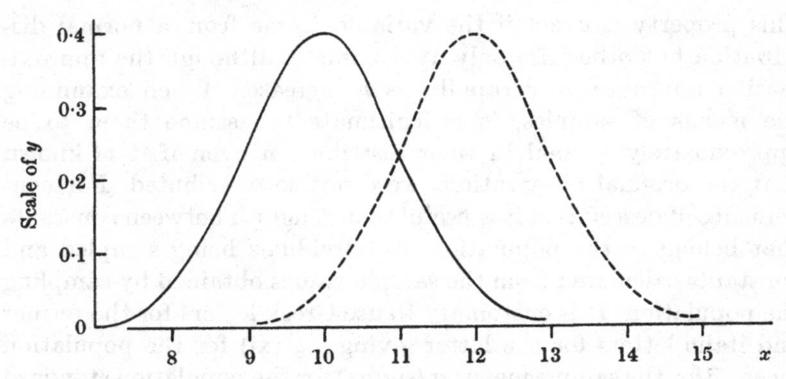

Fig. 11.2. Normal distributions $\xi_0 = 10$, $\xi_1 = 12$, $\sigma = 1$ in both cases

means of samples of sixteen from the two populations specified in (i) and (ii) above are:

(i) normal distribution mean 10, standard deviation 0·25 ($= 1/\sqrt{16}$);

(ii) normal distribution mean 12, standard deviation 0·25 ($= 1/\sqrt{16}$).

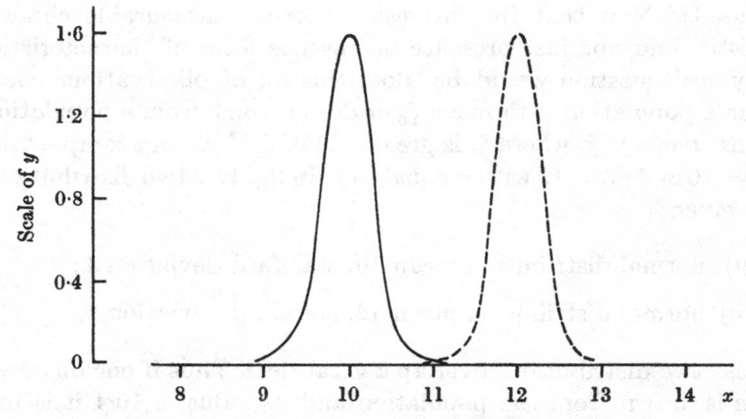

Fig. 11.3. Distributions of means of samples of 16

Fig. 11.3 shows how these two distributions only overlap very slightly and the value 10·4 for the observed mean of sixteen observations is a very strong indication that the population being sampled is the population that has a mean equal to 10.

11.4 For small samples it is not, as a rule, possible to state that the observations come from one population rather than another with absolute certainty. Accordingly the questions asked are framed in more qualitative terms such as 'do the sample results indicate a significant increase in the mean value from its previous value?' To do this an examination is made (as in the previous chapter with the binomial distribution) to see whether the sample results depart significantly from the basic situation. A test of this form is referred to as a *test of significance* and is illustrated below by an example.

Example 11.2 A machine is packaging nominal 8 oz. packets of sugar and it has been found that over a long period the actual weight of sugar put in the packet has been normally distributed with a mean of 8·1 oz. and a standard deviation of 0·04 oz. The setting on the machine which regulates the amount of sugar put in is thought to have been accidentally altered, and to discover whether this is so, a sample of ten packets is examined and the sugar weighed. For the sample the mean weight of sugar in the ten packets is found to be 8·123 oz.

First of all it seems a reasonable assumption that even if the average amount of sugar per packet has been altered the standard deviation of such amounts will be unchanged. It is therefore assumed that the standard deviation of the amounts is 0·04 oz., irrespective of the value of the mean. Since the new level of packaging is unknown all that can be done is to see whether the observed results are consistent with there being no change in the level of packaging. If the results are inconsistent with this, the conclusion would have to be drawn that the level had in fact altered. Now the means of samples of ten from the original population would be distributed normally, with mean 8·1 oz., and standard deviation $0{\cdot}04/\sqrt{10}$. Hence the quantity

$$\frac{\text{Sample mean} - \text{Population mean}}{\text{Standard deviation of sample mean}} = \frac{\bar{x} - 8{\cdot}1}{0{\cdot}04/\sqrt{10}}$$

is a unit normal variable, that is to say it has a mean of zero and a standard deviation of unity. Here $\bar{x}$ is equal to 8·123 and so

$$\frac{\bar{x} - 8{\cdot}1}{0{\cdot}04/\sqrt{10}} = \frac{0{\cdot}023}{0{\cdot}0126} = 1{\cdot}818.$$

From table 9.4 it is found that the area to the right of the ordinate at the abscissa 1·818 of a unit normal variable is equal to 0·0345. Thus in a proportion of only 0·0345 of samples, or about once in thirty times, would a mean as large or larger than that obtained here result from sampling the population with mean 8·1 oz. This is quite an unlikely occurrence and hence it would be reasonable to say that there is some more likely population from which these samples are drawn. If the population mean was higher, 8·12 oz. say, the chances of getting a mean as large as 8·123 or larger goes up to 0·41. This is now a much more likely event and is good evidence that the mean value has in fact increased. The right-hand or upper tail of the normal distribution was used for the test of significance because, if the mean of the population increases, values in the right-hand tail have much higher probabilities.

11.5 One other consideration must, however, be borne in mind. The value 8·123 oz. has been taken as a significant result, significant in the sense in that it shows some departure from the standard weight of 8·1 oz. But suppose that the mean weight of the ten sample packages had come to be 8·077 oz. which is as far below 8·1 as 8·123 is above. In this case the standardised variable is

$$\frac{\bar{x}-8\cdot 1}{0\cdot 04/\sqrt{10}} = -1\cdot 818.$$

From table 9.4 the probability of getting the observed value of $\bar{x}$ or a more extreme value, that is a lower value, is equal to 0·0345. Since this is indicative of an event that is quite rare it is reasonable to conclude that the average level of loading has in fact changed. This shows that getting a mean of 8·123 oz. and saying that it is indicative of a change in population mean, implies also that a mean of 8·077 or less will be taken as indicative of a change in mean. Hence overall the proportion of times that such a divergent or more divergent mean could occur by chance is not 0·0345 but $2 \times 0\cdot 0345$ or 0·069. This is not quite such an unlikely event and in fact would occur about once in every fourteen times, and might not be regarded as so exceptional as to throw doubt on the assumption that the mean is equal to 8·1.

This calculation does not imply that the probability is always automatically doubled whenever the probability of one discrepancy has been calculated. The appropriate probability has to be determined in the light of the question that has been posed. Basically three types of question can be posed and for each type the treatment is slightly different.

(*a*) The machine had an average of 8·1 oz. After the accident is the mean still the same or has it increased?

If the population mean of 8·1 oz. is rejected, the alternative now is that it is greater than 8·1 oz. A sample mean of less than 8·1 oz. is clearly more likely to come from a population with mean 8·1 oz. than from a population whose mean is greater than 8·1 oz. Thus any observed mean less than 8·1 oz. gives no indication of a change in population mean, because even if there is a small probability of such a mean arising from a population whose mean is 8·1 oz. there is an even lower probability of such a mean arising if the population mean is greater than 8·1 oz. The observed mean in the sample of ten was 8·123 oz. and probability theory says that the chance of such a mean, or a greater one, arising, by chance from a population with mean 8·1 oz. is 0·0345 and this is such an unlikely event that it is rejected in favour of the mean being greater than 8·1 oz. The probability is not doubled to include the other tail of the distribution since any observed mean below 8·1 oz. will not in any way suggest that the mean has increased above 8·1 oz.

(*b*) The machine had an average of 8·1 oz. After the accident is the mean still the same or has it decreased?

This situation is just the reverse of (*a*) above. Sample means that are below 8·1 oz. now throw doubt on the mean being unchanged. Those that are above 8·1 oz. are even more unlikely to occur if the population mean has decreased and hence such means are no indication of a change. An observed sample mean of 8·077 oz. implies that there is a probability of 0·0345 that such a result, or a more divergent one, could arise by chance in sampling from a population whose mean is 8·1 oz. Since this is such an unlikely happening it suggests that the mean has in fact decreased, because with a decreased mean the probability of such an observed mean is much greater.

(*c*) The machine had an average of 8·1 oz. After the accident is the mean still the same or has it altered, that is, either increased or decreased?

This is the original problem that was considered. Here a sample mean that is above or below the population mean of 8·1 oz. may be considered significant since such sample means could arise if the mean had in the first case increased and in the second case decreased. This leads to the use of the combined probabilities from the two tails of the distribution, since an observed mean above 8·1 oz. must be just as significant as an observed mean the same amount below 8·1 oz.

Summarising, the three situations lead to slightly different tests:
(*a*) uses the right-hand tail of the distribution;
(*b*) uses the left-hand tail of the distribution;
(*c*) uses both tails of the distribution.

The decision as to which situation is the correct one must be decided from the relevant wording of the particular problem before the numerical analysis is carried out. It is quite incorrect to carry out the analysis and then choose whichever of the three cases produces the most significant result. This will lead to erroneous decisions and the only safe course is to decide in advance which is the appropriate situation.

11.6 When a test of significance has been carried out the result is given in the form of a probability. This probability states the odds of such an observed result occurring by chance when the original situation is true. As this probability gets smaller and smaller there comes a time when it is felt that the odds are so long that another alternative situation must be the true one. In the foregoing it was suggested that although a probability of 0·0345 was small enough for the basic situation to be rejected, a probability of 0·069 was not small enough. It is impossible to lay down hard and fast rules as to the exact levels where one situation is rejected in favour of another, as so much depends on the background of the problem under discussion. If making a wrong decision is only a small matter that subsequent experiments can put right at little or no cost to the manufacturer or consumer then a probability of, say, 0·05 might be appropriate. If, on the other hand, a decision to change involves costly new plant and equipment the change should not take place unless it seems certain that there is a difference, and so a lower level of probability is required, say, 0·01 or even 0·001. There has grown up a rule of thumb which states:

if probability greater than 0·05: take no action;

if probability between 0·05 and 0·01: a warning, further observations desirable;

If probability less than 0·01: take action.

A probability lower than 0·05 is frequently marked with one asterisk (*) and one lower than 0·01 with two asterisks (**). This is conventional only and it does not mean that every experiment should automatically be judged on these standards. Each experiment should be judged in the light of all the background information available and a decision made on the appropriate probability that is to be regarded as significant. The probability deduced from a test of significance is referred to as the *level of significance*. Thus when the probability corresponding to some test is 0·05 this can be referred to as the 5 per cent level of significance.

The tests of significance ultimately require translation of a unit normal deviate into a probability. Since the deviates rarely turn out to be simple quantities this entails using table 9.4 to obtain probabilities for values of x between those tabulated. This can be avoided to some extent by tabulating the values of the abscissa, x, corresponding to specified tail probabilities. In table 11.1 some values are given corresponding, in the first place, to the one-tailed test. Thus the probabilities tabulated are for a normal variable to be 'below x_l' or 'above x_u' separately but not for the combined event 'either below x_l or above x_u'. For the latter event the tabulated probabilities have to be doubled and these values are given in the last column of table 11.1.

Table 11.1. *Normal distribution probabilities*

x_l	x_u	Probability of being 'below x_l' or 'above x_u'	Probability of being 'either below x_l or above x_u'
−1·2816	+1·2816	0·1	0·2
−1·6449	+1·6449	0·05	0·1
−1·9600	+1·9600	0·025	0·05
−2·3263	+2·3263	0·01	0·02
−2·5758	+2·5758	0·005	0·01
−3·0902	+3·0902	0·001	0·002

11.7 *Example* 11.3 Over a period of time a sample of coal has been taken from each successive truck-load of coal delivered from a colliery and it is found that the measurements of percentage ash content of the coal have a mean value of 14·2 and a standard deviation of 1·2. A second colliery now provides coal of a supposedly similar quality, and a sample of five observations from successive trucks is taken and analysed giving percentage ash contents of

13·8, 15·1, 14·6, 15·8, 14·7, respectively.

The question to be answered is whether the percentage ash content in the coal from the second colliery is the same as from the first colliery.

The mean of the five sample measurements is 14·8 and this has to be compared with the previous mean of 14·2. There is thus a difference of $+0·6$ and this difference has to be examined to see whether or not it could be regarded as significant. The test used will be a two-tailed test since the question merely asks whether there could be any difference and not whether there is a difference in any particular direction. It will be assumed that the standard deviation of 1·2 is unchanged. This seems reasonable enough, and the five sample values are consistent with their having a standard deviation of 1·2. Hence the mean of a sample of five will have a standard deviation of $1·2/\sqrt{5}$ and the criterion to be used for the test of significance is

$$\frac{14·8-14·2}{1·2/\sqrt{5}} = 1·118.$$

From table 9.4 the area beyond the ordinate at 1·118 for a unit normal distribution is 0·132, and hence the area in both tails combined is equal to $2 \times 0·132$ or 0·264. Thus such a result, or an even more divergent one would occur in about a quarter of experiments where five samples were drawn from a population with a mean value of 14·2. It seems reasonable to conclude that there is no evidence of a change in the mean value of the ash content of the coal produced.

Example 11.4 It has been found that the length of cuckoos' eggs in place A is distributed with a mean of 22·30 mm. and a standard deviation of 0·9642 mm. Whilst visiting place B an ornithologist collects fifty-eight eggs of the same variety of cuckoo, examines them and measures their length. The mean of the fifty-eight lengths is 22·61 mm. Assuming that the standard deviation of the length of egg is the same in the two places, does this evidence show that the eggs of the cuckoo are longer in place B than in place A?

First of all there is no difference between the eggs in the two places the mean of a sample of fifty-eight eggs should have a normal distribution with mean 22·30 mm. and a standard deviation of $0{\cdot}9642/\sqrt{58}$ mm. Thus the quantity

$$\frac{\bar{x}-22{\cdot}30}{0{\cdot}9642/\sqrt{58}}$$

should be a unit normal variable. In this case $\bar{x}$ is equal to 22·61 mm. and so the quantity reduces to

$$\frac{22{\cdot}61-22{\cdot}30}{0{\cdot}9642/\sqrt{58}} = 2{\cdot}449.$$

Since the question asked is whether those found at B are longer or not, the right-hand or upper tail is needed for a test of significance.

Reference to table 9.4 shows that the area to the right of the ordinate at 2·449 of a unit normal distribution is equal to 0·0072. Hence only about once in every 140 times would such a divergent result be produced with a sample of fifty-eight from a population with a mean of 22·30 mm. It seems much more likely, therefore, that the true mean is greater than 22·30 mm. since if that were so the sample result would have a much greater probability of occurrence. Thus the hypothesis that the length of eggs at place B is the same as at place A would be discarded in favour of the theory that they are longer at place B.

11.8 So far the tests have all examined whether, on the basis of a sample of observations, the mean could be equal to a stipulated value. Suppose now the problem is slightly different, and instead of comparing the mean in one sample with some standard amount, the means in two samples are compared with each other. In this case the basic situation is that the two samples have come from the same population, so that the two means should not differ at all. Due to sampling fluctuations some difference between the means is observed, and the question arises as to how large that difference can be before it becomes significant of a difference between the means in the two populations from which the samples have been drawn. Let $\bar{x}_1$ be the mean of the sample of n_1 individuals drawn from the first population and let the standard deviation of individuals in the population be σ_1. Similarly for the sample from the second population the mean is taken as $\bar{x}_2$, the number of individuals as n_2 and the standard deviation of the population as σ_2. The quantity that has to be examined is the difference

$$\bar{x}_1-\bar{x}_2,$$

and it can be shown that if the two populations from which the samples have been randomly drawn have the same mean, then this quantity is approximately normally distributed with a mean value of zero and a standard deviation of

$$\sqrt{\left(\frac{\sigma_1^2}{n_1}+\frac{\sigma_2^2}{n_2}\right)}.$$

This result is exact if the original populations are both normal but is otherwise only approximate, although as n_1 and n_2 increase it becomes more and more nearly exact. The following example illustrates the whole procedure.

Example 11.5 Two plantations in Malaya are supplying rubber in batches to a factory. The factory has over a period of years been checking the tensile strength in kilograms per sq. cm. on samples from each plantation and has found that the standard deviations of the tensile strengths are 6 kg. for the first plantation and 8 kg. for the second plantation. The factory is interested to know whether the mean tensile strength of the rubber in a certain batch is the same for rubber from the two plantations and examines twelve specimens from the first plantation and sixteen from the second with the following results (in kilograms per sq. cm.):

First plantation	201	201	181	193	179	183		
	188	182	197	185	204	198		
Second plantation	183	189	201	174	194	169	181	199
	178	174	198	188	196	171	195	170

From these figures the following values are obtained

$$\bar{x}_1 = 191, \quad \bar{x}_2 = 185;$$

$$\sqrt{\left(\frac{\sigma_1^2}{n_1}+\frac{\sigma_2^2}{n_2}\right)} = \sqrt{\left(\frac{36}{12}+\frac{64}{16}\right)} = \sqrt{7}$$

$$= 2{\cdot}6458.$$

If there is no difference between the means of the two populations, the quantity $\bar{x}_1-\bar{x}_2$ will have zero mean and standard deviation of

$$\sqrt{\left(\frac{\sigma_1^2}{n_1}+\frac{\sigma_2^2}{n_2}\right)}$$

which implies that $$\frac{\bar{x}_1-\bar{x}_2}{\sqrt{\left(\dfrac{\sigma_1^2}{n_1}+\dfrac{\sigma_2^2}{n_2}\right)}}$$

will be a unit normal variable. In this example

$$\frac{\bar{x}_1 - \bar{x}_2}{\sqrt{\left(\frac{\sigma_1^2}{n_1} + \frac{\sigma_2^2}{n_2}\right)}} = \frac{6}{2{\cdot}6458} = 2{\cdot}268.$$

From table 9.4 the area of the normal distribution to the right of the ordinate at 2·268 is equal to 0·0117. However, a two-tailed test is required here, since the question asks whether the strength is the same in the two plantations and not whether it is greater in one particular plantation. This gives the appropriate probability as $2 \times 0{\cdot}0117$ or 0·0234. Thus this particular result would be rare if the two means were the same, but perhaps not quite uncommon enough to be conclusive, and hence although the evidence is highly suggestive that the first plantation has a greater strength it is not overwhelmingly strong and some more observations would be desirable.

11.9 The two tests that have been used in this chapter can be summarised as follows:

One sample test. Sample of n. Mean of sample $\bar{x}$.

Population mean ξ. Population standard deviation σ.

Decide whether left-tail, right-tail or two-tailed test.

Calculate

$$u = \frac{\bar{x} - \xi}{\sigma/\sqrt{n}}.$$

Refer to normal distribution table for corresponding probability.

Find appropriate significance level.

Two sample test. First sample, mean $\bar{x}_1$, number n_1. Second sample, mean $\bar{x}_2$, number n_2.

Standard deviations of the two populations σ_1 and σ_2 respectively.

Decide whether left-tail, right-tail or two-tailed test.

Calculate

$$u = \frac{\bar{x}_1 - \bar{x}_2}{\sqrt{\left(\frac{\sigma_1^2}{n_1} + \frac{\sigma_2^2}{n_2}\right)}}.$$

Refer to normal distribution table for corresponding probability.

Find appropriate significance level.

EXERCISES

11.1 An analyst is making repeated determinations of the percentage fibre in soya cotton cake and has found that the standard deviation of determinations from the same batch of cake is 0·12, and that the mean of the determinations is 12·40. A new batch of cotton cake arrives and from it ten samples are taken and analysed giving

12·46, 12·30, 12·43, 12·41, 12·58, 12·37, 12·63, 12·25, 12·37, 12·48.

Use these figures to test whether the mean percentage fibre is still equal to 12·40, or whether it has increased in the new batch.

11.2 The heights of men in a certain large town are normally distributed and have a mean of 68·40 in. and a standard deviation of 2·13 in. Find the probability

(*a*) that a man selected at random has a height over 6 ft.;
(*b*) that the mean of a sample of 120 men is less than 68·34 in.

11.3 Steel rods are usually being manufactured with a mean length of 14·50 cm. and a standard deviation of 0·73 cm. One batch has, however, been manufactured with a mean length of 14·30 cm., the standard deviation being unchanged. If a sample of ninety-five rods are taken and measured from each batch is it reasonable to say that the batch which has a length of only 14·30 cm. will be easily detectable?

11.4 Specimens of a certain type of string have a mean breaking strength of 17·1 lb. and a standard deviation of 1·8 lb., the breaking strengths being approximately normally distributed. A new method of manufacture is tried in order to increase the breaking strength of the string. A sample of twelve pieces is taken, giving the following breaking strengths

16·6, 19·3, 17·9, 18·7, 14·7, 18·0, 19·9, 20·6, 16·9, 17·3, 15·7, 16·8.

Do these results indicate any significant improvement in breaking strength?

11.5 Measurements were made on a large number of terminal leaflets of a variety of strawberry plants grown in open borders, and it was found that the mean leaflet area was 21 sq. cm. and the standard deviation 3 sq. cm. Some plants of the same variety were also grown in a greenhouse, and from these, nine were selected at random and their terminal leaflet areas in sq. cm. were

27, 24, 22, 23, 18, 26, 20, 19, 22.

Do these figures suggest that the mean terminal area of plants grown in pots in a greenhouse is any different from that of plants grown in the open?

11.6 To investigate the movement of antibiotics in broad bean plants the plant is treated for 18 hr. with a solution of chloramphenicol and at the end the concentration determined. Experiments on rooted plants have given a mean concentration of 54·3 (milligrams per gram of fresh weight) with a standard deviation of 4·5. Ten cut shoots of broad bean plants are now treated in the same way and the concentrations are

50, 53, 58, 57, 63, 62, 55, 65, 46, 60.

Do these figures indicate any change in the mean level of concentration of chloramphenicol? Assume that the standard deviation is unaltered.

11.7 The lengths of eggs of the common tern are approximately normally distributed with a mean of 4·11 cm. and a standard deviation of 0·19 cm. A sample of eight eggs was collected from a completely fresh part of the coast and the lengths, in centimetres, where

4·1, 4·4, 4·5, 4·1, 3·9, 4·4, 4·6, 4·5.

Would you regard this sample as indicating a real difference in egg-length in the new locality?

11.8 Cement mortar briquettes are being made and the breaking strength of the briquettes measured. The standard deviation of the breaking strength has been found to be 17 lb. Two samples each of ten briquettes are available and the breaking strengths are

Sample *A*	518	508	554	555	536	544	532	530	554	542
Sample *B*	544	538	554	540	506	534	548	530	525	522

Test whether there is any difference between the mean strength obtained in the two samples.

11.9 Two varieties of tomato are grown. Previous experience has shown that the standard deviation of the yield in kilograms from plants of either variety is equal to 0·32.

Ten plants of variety *A* and eight of variety *B* are harvested with the following results:

Variety *A*	1·375	1·407	1·068	1·752	1·773	1·201	0·779	1·042	1·223	1·633
Variety *B*	1·033	1·217	0·984	1·615	1·693	0·673	0·840	1·252		

Test whether the yield of the two varieties of tomato is the same.

11.10 Two different makes of tyre were used on a car and the wear in thousandths of an inch after 1000 miles travelling measured. Twenty-four tyres of make *A* and twelve tyres of make *B* were used in the experiment, the results being as follows:

Tyre *A*	13·4	17·3	23·3	13·4	13·7	18·5	26·9	15·3
	19·0	18·0	21·6	15·8	14·0	17·6	22·2	14·3
	15·0	16·4	23·8	14·9	18·4	17·9	24·7	18·2
Tyre *B*	11·9	13·4	16·2	15·6	17·9	18·2	11·3	13·1
	12·2	15·0	17·3	17·0				

Previous experiments have shown that the standard deviations of the wear per 1000 miles are 3·2 for tyre *A* and 2·1 for tyre *B*. Investigate whether there are any differences in the average wear of the two makes of tyre.

11.11 The diameter in millimetres of ears of wheat is measured for two samples, *A* and *B*. In sample *A*, fifty-one ears are measured whilst in *B*, sixty-one ears are measured. The results are given below:

	Diameter (central values)							
	38·5	40·5	42·5	44·5	46·5	48·5	50·5	Total
No. in sample *A*	—	5	16	19	8	1	2	51
No. in sample *B*	1	5	15	26	11	3	—	61

From previous experience it has been found that the standard deviation of the diameter of ears of wheat is 1·84 mm. Use this information to test whether the means in the two samples are equal.

11.12 A sample of private houses in two large towns was chosen from the rating list. Interviewers were sent to the houses to inquire if the occupants possessed a television set and if so how much it cost. The results were as given in the following table:

	No. of television sets	Average cost (£)	Standard deviation of cost (£)
Town *A*	14	71	10
Town *B*	20	65	15

Assuming the standard deviations for the whole towns are given by the standard deviations observed in the samples, test whether town *A* tends to buy more expensive sets than town *B*.

11.13 A voltmeter is being used to compare the voltage of two so-called standard cells. It has been found in the past that if a series of independent readings of a standard cell are made they are normally distributed about the true voltage with a standard deviation of 0·025 volts. Seven readings were made on each of the two cells. The first cell gave a mean reading of 1·172 volts and the second of 1·143 volts. Do you consider that the two standard cells are giving the same voltage or not?

11.14 The mean and standard deviation of the weights of 18-year-old men recruited for the Army in one year were 137·3 and 31·1 lb. respectively. Fifteen men were selected for a special duty and their weights were (in lb.):

104·2, 173·2, 141·5, 172·9, 201·7, 203·1, 171·8, 122·5,
167·5, 181·4, 177·7, 176·7, 161·0, 158·2, 180·3.

Do you consider that the men selected were a random sample from the population of Army recruits?

11.15 The purity of a chemical manufactured on a large scale varies slightly from batch to batch. Over the past the purity has had a mean value of 68·4 per cent and a standard deviation of 2·3 per cent. A small modification of the manufacturing process is now made and the purity of the first eleven batches produced is:

66·1, 71·3, 75·2, 64·3, 76·4, 75·6, 66·3, 63·2, 65·8, 62·4, 73·4.

Assuming that the standard deviation is unchanged, has the modification improved the process?

12

FURTHER TESTS OF SIGNIFICANCE

12.1 The previous chapter was concerned with tests designed to determine, on the basis of a sample, whether the population mean was equal to some specified amount. An alternative test is required when it is desired to investigate whether the variability of the observations in a population has some specified value, on the basis of a sample of n observations from the population. Thus a local corporation might find it more economical to change all its street-lamp bulbs at a fixed time instead of waiting for individual complaints that a lamp has gone out and then sending a man to replace it. If this policy is to be successful, however, the variability of the bulbs must be small, because it would be wasteful to change all the bulbs if only a few were burnt out and many burning hours left in the rest. On the other hand, to delay changing them would result in angry protests from the residents in the streets concerned. Hence the corporation would require bulbs that have little variation about the nominal length of life. A sample of bulbs could be examined from time to time by burning them to extinction in order to see whether the variability was remaining constant.

Suppose that in the past the average length of life of the bulbs has turned out to be 1,608 burning-hourswith a standard deviation of 381 hr. A check sample of the bulbs delivered one month is examined and the eight bulbs selected gave lengths of life (in hours) of

1507, 1982, 1587, 1221, 893, 1818, 2029, 2147.

The mean of these eight observations is 1,648, and this does not seem to be very far removed from the nominal mean of 1,608 hr. If anything there is an improvement in length of life which would no doubt be welcome to the corporation concerned. Next the variability is examined and at first sight there seems to be quite a large variation in the sample values. Calculations give

$$\Sigma(x-\bar{x})^2 = 1{,}310{,}574, \quad s^2 = \frac{1}{n}\Sigma(x-\bar{x})^2 = 163{,}821{\cdot}75,$$

$$s = 404{\cdot}75 \text{ hr.},$$

and the first impression given by the figures is that there is a somewhat greater variability amongst the bulbs than that laid down in the original standard. However, it is to be expected that if samples are drawn from a population with a standard deviation of 381 hr., some of those samples will have standard deviations greater than 381 hr. and some of them lower than 381 hr. The form of the deviations of the sample standard deviations from 381 is, however, no longer in the fashion of a normal distribution as it was for the sample mean. Another basic distribution has to be introduced here and stated formally, it reads:

A sample of n observations is randomly drawn from a normal distribution whose standard deviation is σ. Then the quantity

$$\frac{ns^2}{\sigma^2} = \frac{\Sigma(x-\bar{x})^2}{\sigma^2},$$

where x represents a sample value, is distributed in the form of a χ^2 distribution.

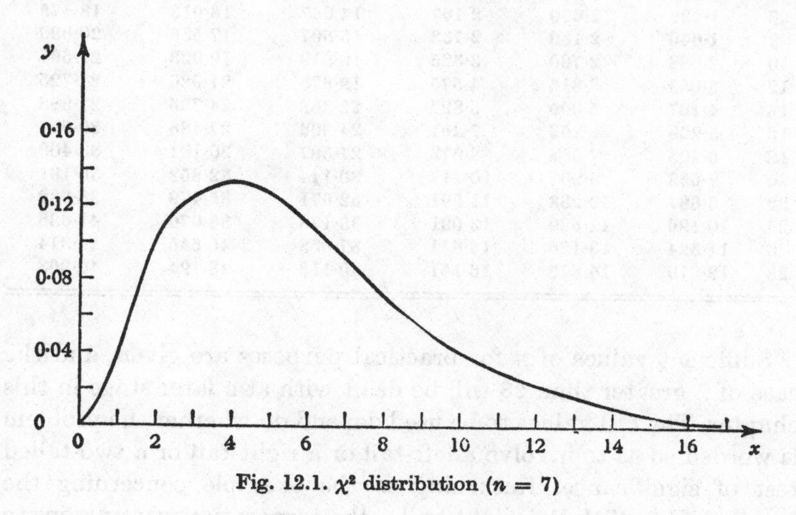

Fig. 12.1. χ^2 distribution ($n = 7$)

A typical χ^2 distribution is shown in fig. 12.1 above, where n is equal to 7, and it will be noticed that the distribution is no longer symmetrical but is rather skew, having a long tail to the right. If repeated samples of size 7 were drawn from a normal population with standard deviation σ and $\Sigma(x-\bar{x})^2/\sigma^2$ calculated for each sample, the values could be grouped and made into a frequency distribution which would look more and more like fig. 12.1 as the number of samples increased. The shape of the distribution varies according to the value of n, and as n increases it gets more and more like a normal curve in appearance. In table 12.1 below the significance levels of the χ^2 distribution are tabulated. Both tails of the distribution are tabulated, columns (2)–(4) giving the abscissae for which the area to the left is the amount stated at the head of the columns, whilst in columns (5)–(7) it is the area to the right.

Table 12.1. *Significance levels of χ^2 distribution*

(1)	(2)	(3)	(4)	(5)	(6)	(7)
	Area up to abscissa equal to			Area beyond abscissa equal to		
n	0·01	0·025	0·05	0·05	0·025	0·01
4	0·115	0·216	0·352	7·815	9·348	11·345
5	0·297	0·484	0·711	9·488	11·143	13·277
6	0·554	0·831	1·145	11·071	12·833	15·086
7	0·872	1·237	1·635	12·592	14·449	16·812
8	1·239	1·690	2·167	14·067	16·013	18·475
9	1·646	2·180	2·733	15·507	17·535	20·090
10	2·088	2·700	3·325	16·919	19·023	21·666
12	3·053	3·816	4·575	19·675	21·920	24·725
14	4·107	5·009	5·892	22·362	24·736	27·688
16	5·229	6·262	7·261	24·996	27·488	30·578
18	6·408	7·564	8·672	27·587	30·191	33·409
20	7·633	8·907	10·117	30·144	32·852	36·191
22	8·897	10·283	11·591	32·671	35·479	38·932
24	10·196	11·689	13·091	35·173	38·076	41·638
26	11·524	13·120	14·611	37·653	40·646	44·314
28	12·879	14·573	16·151	40·113	43·194	46·963

Sufficient values of n for practical purposes are given, and the case of n greater than 28 will be dealt with at a later stage in this chapter. The tail values to be used depend on whether the problem is worded so as to involve a left-tail or a right-tail or a two-tailed test of significance. Returning to the example concerning the lengths of life of electric light bulbs, the corporation was anxious to avoid excessive variability, and hence the alternative to be considered is that the standard deviation is in excess of 381 hr. The appropriate test will then be the right-tailed test. The figures obtained for the sample give

$$\frac{\Sigma(x-\bar{x})^2}{\sigma^2} = \frac{1,301,574}{145,161} = 9{\cdot}03.$$

Reference to table 12.1 ($n = 8$) shows that this value is quite a long way short of the 0·05 significance level which is 14·07, and hence on the evidence of the sample there is no reason to doubt that the variability of the bulbs is unchanged at 381 hr.

12.2 Two further examples of tests of significance that use the χ^2 tables will now be given.

Example 12.1 An experiment was performed to determine the strength of resin films. First a series of metal panels of standard thickness were coated with a resin film whose thickness was known. The strength of the resin film was then measured by a machine which pressed a steel ball into the panel. This had the effect of stretching the film on the reverse side until it broke, when an electric contact was completed through the panel and the machine stopped. The strength of the film was taken to be proportional to the depth of penetration of the steel ball.

It was rather important that the variability of the strength of the resin films should be kept as low as possible. The standard deviation of the films was equal to 4·7 units. By introducing a slightly modified manufacturing technique, it was hoped to reduce the standard deviation still further. From a sample of seven films, the strengths obtained were

79·3, 72·1, 80·6, 77·1, 75·0, 73·9, 70·1.

Using these values it is desired to investigate whether the variability has in fact decreased. Now

$$\bar{x} = 75{\cdot}443 \quad \text{and} \quad \Sigma(x-\bar{x})^2 = 86{\cdot}52.$$

Hence
$$\frac{\Sigma(x-\bar{x})^2}{\sigma^2} = 3{\cdot}92.$$

From the values in table 12.1 corresponding to $n = 7$, this value is seen not to be significant, remembering that judgment is being based on the left-hand tail of the distribution. The value 3·92 is quite definitely below the mean of the distribution, and it may well be that a further sample would demonstrate the significance of a different and reduced variability.

Example 12.2 The tensile strength, in kg./cm.^{2}, of specimens of rubber from one plantation was measured. Past experience had shown that the standard deviation of the tensile strength of specimens from this plantation was equal to 13·7 kg./cm.². It was very important from the commercial point of view that the strength should not have a great variability, and when a random sample of six specimens was taken from a new consignment, the following tensile strengths in kg./cm.² were obtained:

177, 173, 137, 196, 145, 168.

To test whether the variability has increased or not the following quantities are calculated:

$$\bar{x} = 166, \quad \Sigma(x-\bar{x})^2 = 2356, \quad \frac{\Sigma(x-\bar{x})^2}{\sigma^2} = 12{\cdot}55.$$

Using table 12.1 ($n = 6$), and considering the right-hand tail, the 0·05 probability level corresponds to 11·07 and the 0·025 probability level to 12·83. The value 12·55 obtained here is thus near to the 0·025 probability level, and seems to be a strong, though not conclusive, indication that there is some increase in the variability.

12.3 Since n can take any integral value it is impossible to construct a table such as table 12.1 for all the values of n and some short cut is needed. As the frequency curve for χ^2 is drawn for larger and larger values of n, it is found that it becomes more and more symmetrical in form and approaches the shape of a normal distribution. This very useful property implies that, if n is large, a test for the variability in the sample can be based on the normal distribution. The latter is well tabulated and its properties of shape do not depend on either its mean or variance.

The procedure adopted consists first of estimating the variability in the population from that in the sample, and then carrying out a test of significance to determine whether or not it differs significantly from some value. Now whilst it is true that the sample mean, $\bar{x}$, gives a good estimate of the population mean, ξ, it is not true that the best estimate of the population standard deviation, σ, will be given by the sample standard deviation, namely,

$$s = \sqrt{\left\{\frac{1}{n}\Sigma(x-\bar{x})^2\right\}}.$$

In fact a better estimate of σ is given by the formula

$$s_1 = \sqrt{\left\{\frac{1}{n-1}\Sigma(x-\bar{x})^2\right\}},$$

where the divisor is now $n-1$ and not n as suggested in section 7.8. Note that there is the relation

$$s^2 = \frac{n-1}{n} s_1^2$$

between the two quantities. This modification of s is important and it must be remembered that whenever the variability in some population is to be estimated on the basis of a sample, the quantity to be used should be s_1 and not s.

Next, the distribution of s_1 is required if s_1 is to be tested for significance. Statistical theory shows that in large samples of n from a normal population, the quantity s_1 is very nearly normally distributed with

$$\text{Mean } (s_1) = \sigma\left(1-\frac{1}{4n}\right), \quad \text{Standard deviation of } s_1 = \frac{\sigma}{\sqrt{(2n)}}.$$

With the help of these quantities the appropriate significance test can now be carried out.

Example 12.3 The vitamin C content, in milligrams per 100 gm. of forty specimens of tomato juice was found to be:

22	17	18	29	17	22	16	23	25	19
21	16	22	21	20	23	22	17	15	13
20	24	24	15	23	16	30	17	18	21
14	14	18	21	24	19	22	21	17	20

It is desired to test whether the variability, or standard deviation, was equal to 3 mg. as in the past or whether an increase in the variability had taken place. Calculations from the data give

$$\bar{x} = 19{\cdot}9, \quad \Sigma x^2 = 16{,}424,$$

$$\frac{1}{n-1}\Sigma(x-\bar{x})^2 = 14{\cdot}9641 \quad \text{and} \quad s_1 = 3{\cdot}8683.$$

Now if no increase in variability has taken place the quantity s_1 in samples of forty should be normally distributed with

$$\text{Mean } (s_1) = 3(1-\tfrac{1}{160}) = 2{\cdot}98125,$$

$$\text{Standard deviation of } s_1 = \frac{3}{\sqrt{80}} = 0{\cdot}3354.$$

Hence the appropriate quantity is

$$\frac{s_1 - \text{Mean } (s_1)}{\text{S.D.}(s_1)} = \frac{3{\cdot}8683-2{\cdot}9813}{0{\cdot}3354} = 2{\cdot}645.$$

Reference to the tables of the normal distribution shows that the chance of as large or a larger value of s_1 being observed in a sample from a population whose standard deviation is 3, is equal to 0·0041. This is very small indeed and is an indication that the variability of the vitamin C content in the specimens of tomato juice has increased.

12.4 The tests evolved in chapter 9 were tests of significance for the situation in which each individual either did or did not possess some property. The expected number of successes was compared with the theoretical number that could be expected if the concept of binomial sampling was a valid one. A further test of agreement between theory and sampling can often be carried out, taking account of the properties of the binomial distribution. First, notice that if a sample of n is drawn from a population in which a proportion p of individuals possess some property, and x are found to have the property, then

$$\text{Mean } x = np, \quad \text{Standard deviation of } x = \sqrt{(npq)}, \text{ where } q = 1-p.$$

If instead of the number x, the proportion x/n is considered, then

$$\text{Mean } \frac{x}{n} = p, \quad \text{Standard deviation of } \frac{x}{n} = \sqrt{\left(\frac{pq}{n}\right)}.$$

Thus, in repeated sampling, the mean value of the proportion of individuals possessing the character in the sample is equal to the proportion in the population being sampled. However, the accuracy of that estimate increases as the number in the sample increases since the standard deviation is $\sqrt{\left(\frac{pq}{n}\right)}$, and for a constant value of p this decreases as n increases. If a series of experiments are made the best estimate of p will be a combined one. Suppose that it is desired to estimate what proportion of beads in a large tub are blue in colour. A small scoop is used and produces the following results:

First attempt	24 beads drawn,	7 blue
Second attempt	28 beads drawn,	10 blue
Third attempt	21 beads drawn,	5 blue
Fourth attempt	25 beads drawn,	6 blue

Since they are four independent attempts or estimates, assuming that after each attempt the beads are returned and well mixed before the next batch is selected, the best estimate would be the overall proportion of blue beads obtained, that is

$$\frac{7+10+5+6}{24+28+21+25} = \frac{28}{98} = 0{\cdot}2857.$$

The standard deviation of this quantity will be

$$\sqrt{\frac{pq}{n}} = \sqrt{\left[\left(\frac{28}{98}\right)\left(\frac{70}{98}\right)\left(\frac{1}{98}\right)\right]} = 0{\cdot}0456.$$

This standard deviation is a measure of the variation that might be expected amongst the estimates of p if the experiment were repeated a large number of times. Provided n is large this standard deviation can be used to test for the significance of an observed proportion. Thus suppose it was desired to test whether the proportion of blue beads in the experiment just described was equal to $\frac{1}{4}$. Then the quantity x/n, where x is the observed number of blue beads in a sample of n, will be approximately normally distributed with mean p, or $\frac{1}{4}$, and standard deviation $\sqrt{(pq/n)}$, where $q = 1-p$. Hence

$$\frac{x/n-p}{\sqrt{(pq/n)}} = \frac{0{\cdot}2857-0{\cdot}25}{0{\cdot}0437} = 0{\cdot}816,$$

and from the tables of the normal distribution this is not significant.

12.5 An extension of the above test occurs when it is desired to test for the equality of the proportions in two populations. Thus suppose the incidence of a particular blood group in two races is being investigated. A sample of n_1 individuals is drawn randomly from the first population, in which the proportion possessing the blood group is p_1 (unknown), and of the n_1 individuals k_1 are found to possess the particular blood group. Another sample, of n_2 individuals, is drawn from the second population in which a proportion p_2 (also unknown) possess the blood group, and k_2 are found to have the blood group. A test is required to investigate whether the two population proportions are the same, that is, whether $p_1 = p_2$, even though this common value remains unknown.

If the two samples are drawn from populations in which there is a common proportion, p, of individuals that possess the character, then the quantity

$$\left(\frac{k_1}{n_1}-\frac{k_2}{n_2}\right)$$

will have, in repeated sampling, a mean value of zero and a standard deviation of

$$\sqrt{\left(\frac{\sigma_1^2}{n_1}+\frac{\sigma_2^2}{n_2}\right)},$$

where σ_1^2/n_1 is the variance of k_1/n_1 and σ_2^2/n_2 is the variance of k_2/n_2. This result was quoted earlier in section 11.8. Now the variance of k_1 is n_1pq and hence of k_1/n_1 is pq/n_1, Hence the standard deviation of $(k_1/n_1-k_2/n_2)$ is

$$\sqrt{\left[pq\left(\frac{1}{n_1}+\frac{1}{n_2}\right)\right]},$$

where $q = 1-p$.

The test criterion to be calculated will thus be

$$\frac{\frac{k_1}{n_1}-\frac{k_2}{n_2}}{\sqrt{\left[pq\left(\frac{1}{n_1}+\frac{1}{n_2}\right)\right]}},$$

and, provided p_1 and p_2 are in fact equal, this quantity will be distributed as a unit normal variable when n_1 and n_2 are reasonably large. To use the test requires, however, a knowledge of p and this is not usually available. This implies that some form of estimate must be substituted for p in order to be able to carry out the test. Since the test is designed to investigate a common proportion the best estimate of p would be found from pooling the samples and taking the overall proportion. Thus take as an estimate of p the quantity

$$\frac{k_1+k_2}{n_1+n_2}$$

giving the modified criterion

$$\frac{\left(\frac{k_1}{n_1}-\frac{k_2}{n_2}\right)}{\sqrt{\left[\left(\frac{k_1+k_2}{n_1+n_2}\right)\left(1-\frac{k_1+k_2}{n_1+n_2}\right)\left(\frac{1}{n_1}+\frac{1}{n_2}\right)\right]}}.$$

Example 12.4 To investigate whether the proportion of people having a certain blood group is the same in two populations a sample of 108 individuals is drawn from the first population and a sample of eighty-one from the second. Of these individuals twenty-five from the first and thirty-five from the second sample were found to have the blood group concerned. Can the proportion be considered equal in the two populations from which these samples were drawn?

The data can be expressed very neatly in tabular form as follows:

	With blood group	Without blood group	Total
First sample	25	83	108
Second sample	35	46	81
Total	60	129	189

The test criterion above gives

$$\frac{\frac{25}{108}-\frac{35}{81}}{\sqrt{\left[\left(\frac{60}{189}\right)\left(\frac{129}{189}\right)\left(\frac{1}{108}+\frac{1}{81}\right)\right]}} = -\frac{0{\cdot}20062}{0{\cdot}06842} = -2{\cdot}932.$$

This must be a two-tailed test since the question is merely designed to investigate whether the proportions are equal or not. From the normal curve tables the area of the left tail beyond $-2{\cdot}932$ is equal to 0·0017 so that the area of both tails would be equal to 0·0034. This probability is extremely small and would seem to indicate that there is not an equal proportion of persons with that particular blood group in the two populations.

Example 12.5 An experiment is carried out with willow cuttings in order to determine whether the number of buds on a cutting affects the proportion of buds which grow after a certain period. A number of cuttings with three buds only are kept in water, and after the given period it is found that twenty-one out of twenty-seven possible buds are growing. At the same time a number of cuttings with six buds are kept and at the end of the same period twenty-one out of thirty-six possible buds are growing. On the basis of the data it is desired to investigate whether the proportion is the same in the two cases.

Put into tabular form the data become:

	Growing	Not growing	Total
Buds (3-bud cuttings)	21	6	27
Buds (6-bud cuttings)	21	15	36
Total	42	21	63

The estimate of a common p is $\frac{42}{63} = \frac{2}{3}$, and the test criterion gives

$$\frac{\dfrac{21}{27}-\dfrac{21}{36}}{\sqrt{\left[\left(\dfrac{2}{3}\right)\left(\dfrac{1}{3}\right)\left(\dfrac{1}{27}+\dfrac{1}{36}\right)\right]}} = \frac{0\cdot 19444}{0\cdot 12001} = 1\cdot 620.$$

Since the alternative to equality is that the proportions may vary either way, this is a two-tailed test, and as such the probability in the two tails for the unit normal curve is $2 \times 0\cdot 053$ or $0\cdot 106$. This probability shows that such a difference as that observed between the sample proportions is not a very unlikely event and, therefore, there is not sufficient evidence to doubt the hypothesis that the two proportions are the same.

12.6 In the last two examples it was necessary to make use of an estimated proportion in order to be able to carry out the required tests. This is often the situation in the binomial type of sampling studied earlier, where it is desired to investigate whether this type of model is in fact consistent with the available data. Consider as an illustration the following example.

Example 12.6 In a duck breeding farm it is believed that the probability of a duck having a white bib is constant and independent of the colour of the bib of any other duck. To test this ninety batches of ducks' eggs were hatched, each batch having five eggs in it. The number of ducks with white bibs in each batch was noted and the following results obtained:

No. with white bibs	0	1	2	3	4	5	Total
No. of batches	27	35	22	4	2	0	90

First assume that the binomial basis is true and that the probability of a duck having a white bib on hatching is constant from duck to duck and that the batches are independent. Then the total number of ducks hatched is 90×5, or 450, and of these 450 ducks the number hatched with a white bib is

$$35 \times 1 + 22 \times 2 + 4 \times 3 + 4 \times 2 = 99.$$

Hence the proportion of ducks with a white bib is 99/450 or $0\cdot 22$. Then if the probability is remaining constant the proportions of occasions when there are 0, 1, 2, 3, 4 or 5, white bibs on the ducks in the batches of five will be given by the successive terms of the binomial expansion

$$(0\cdot78+0\cdot22)^5.$$

Expanding this series the terms are

$$0\cdot2887,\quad 0\cdot4072,\quad 0\cdot2297,\quad 0\cdot0648,\quad 0\cdot0091,\quad 0\cdot0005.$$

Since ninety such sets of five eggs have been observed the expected numbers of each of the six types will be

$$26\cdot0,\quad 36\cdot6,\quad 20\cdot7,\quad 5\cdot8,\quad 0\cdot8,\quad 0\cdot1,$$

and thus a comparison of the observed and expected numbers gives the following table:

No. of white bibs ...	0	1	2	3	4	5	Total
Observed no. of batches	27	35	22	4	2	0	90
Expected no. of batches	26·0	36·6	20·7	5·8	0·8	0·1	90

A comparison of the observed and expected frequencies in the table indicates that there is a very good agreement, and suggests that the hypothesis of a binomial set-up with constant probabilities adequately describes the data.

12.7 In example 11.3 where the mean ash content of successive trucks of coal was compared with a standard, it was tacitly assumed that the standard deviation was the same as that found in earlier analyses. This may not be the case, and a more representative result would be obtained by using the standard deviation estimated from the sample itself. The ash contents were

$$13\cdot8,\quad 15\cdot1,\quad 14\cdot6,\quad 15\cdot8,\quad 14\cdot7$$

with a mean of 14·8 and

$$s_1^2 = \tfrac{1}{4}(1\cdot0^2+0\cdot3^2+0\cdot2^2+1\cdot0^2+0\cdot1^2) = 0\cdot535,$$

$$s_1 = 0\cdot731.$$

Instead of calculating $u = \dfrac{\bar{x}-\xi}{\sigma/\sqrt{n}}$

the appropriate criterion is now

$$t = \frac{\bar{x}-\xi}{s_1/\sqrt{n}}.$$

In this particular case

$$t = \frac{14\cdot8-14\cdot2}{0\cdot731/\sqrt{5}} = 1\cdot837.$$

This criterion now has to be referred to the t-tables given in table 12.2 in place of the normal distribution table previously used for the u criterion. From this table the two-tailed situation with $n = 5$ gives a probability somewhere between 0·1 and 0·2. Thus it still seems reasonable to conclude that there is no evidence of a change in the mean value of the ash content of the coal produced.

Table 12.2. *Significance levels of t-distribution*

Size of sample n	Probability* of a deviation greater than t				
	0·25	0·1	0·05	0·01	0·005
2	1·000	3·078	6·314	31·821	63·657
3	0·816	1·886	2·920	6·965	9·925
4	0·765	1·638	2·353	4·541	5·841
5	0·741	1·533	2·132	3·747	4·604
6	0·727	1·476	2·015	3·365	4·032
7	0·718	1·440	1·943	3·143	3·707
8	0·711	1·415	1·895	2·998	3·499
9	0·706	1·397	1·860	2·896	3·355
10	0·703	1·383	1·833	2·821	3·250
12	0·697	1·363	1·796	2·718	3·106
15	0·692	1·345	1·761	2·624	2·977
20	0·688	1·328	1·729	2·539	2·861
30	0·683	1·311	1·699	2·462	2·756
40	0·681	1·304	1·685	2·426	2·708
50	0·680	1·299	1·678	2·407	2·680
100	0·677	1·291	1·662	2·368	2·631
∞	0·674	1·282	1·645	2·326	2·576

* The probability of a deviation numerically greater than t (i.e. the two-tailed situation) is twice the probability shown at the head of the column.

The necessity to use the t-distribution, rather than the u distribution, arises because in the former an estimate, s, of σ is used whose accuracy depends upon the sample size, n, employed for this purpose. One has to be a little careful in applying it to small samples to ensure that the variable possesses an approximate normal distribution. The large-sample method does not require this precaution since the mean is likely to be very nearly normally distributed. However, it does possess the more serious fault of requiring a knowledge of σ which is unlikely to be available in many sample problems.

EXERCISES

12.1 The lengths of a sample of ten rivets from a large batch are measured in centimetres, and found to be:

1·89, 2·21, 2·03, 1·94, 1·97, 2·13, 2·08, 2·01, 1·95, 2·11.

It is desirable that the standard deviation of the lengths of rivets from the same batch should not exceed 0·065 cm. Test whether the standard appears to have been attained.

12.2 In testing projectiles it is desirable to limit variation in their muzzle velocity as much as possible, in order to prevent variations in their performance. Fifteen projectiles were fired and their muzzle velocities in ft./sec. noted as

1365, 1362, 1351, 1358, 1355, 1354, 1348, 1363,
1352, 1353, 1357, 1361, 1347, 1363, 1362.

In the past the standard deviation of the muzzle velocity has been 7·2 ft./sec. Do the present figures indicate any improvement on that value?

12.3 Raw rubber is treated with chemicals and then subjected to a vulcanising process known as curing. Eight specimens are then subjected to a new process and their moduli of elasticity in kg./cm.^{2} measured. The values obtained are

32·0, 30·9, 33·3, 31·0, 34·3, 32·4, 31·1, 31·0.

In past cases the standard deviation of the modulus of elasticity has been equal to 1·1 kg./cm.^{2}, and it is believed that although the new process has produced a more elastic rubber the standard deviation has increased. Test whether this is so or not.

12.4 The table below gives the intelligence quotient of 101 grammar schoolboys specialising in science. Use the table to test whether the standard deviation of I.Q. was equal to 8·4 or not.

I.Q.	No. of boys	I.Q.	No. of boys
129·5–134·5	1	99·5–104·5	14
124·5–129·5	2	94·5–99·5	12
119·5–124·5	8	89·5–94·5	11
114·5–119·5	11	84·5–89·5	7
109·5–114·5	15	79·5–84·5	3
104·5–109·5	16	74·5–79·5	1
		Total	101

12.5 The heights of sixty Swedish men were obtained and are given below. Test whether the standard deviation of the population from which they were selected was equal to 5·7 cm.

Height (cm.), central values	No. of men	Height (cm.), central values	No. of men
151	1	172	8
154	1	175	9
157	1	178	5
160	4	181	1
163	6	184	2
166	8	187	—
169	13	190	1
		Total	60

12.6 One hundred and forty-seven schoolchildren between the ages of $9\frac{1}{2}$ and 10 years were given extra pasteurised milk for four months and their change in weight (gain or loss) measured. A very large group of schoolchildren of similar age who were not given the milk had a standard deviation of change in weight of 22 oz. Test whether those children given the extra milk had a more variable gain in weight over the four months or not.

Change of weight (oz.), central values	No. of children	Change of weight (oz.), central values	No. of children
−45	2	36	15
−36	2	45	13
−27	5	54	7
−18	13	63	3
−9	14	72	3
0	14	81	1
9	23	90	1
18	17	Total	147
27	14		

12.7 The diameters of ball-bearings are known to be normally distributed, with a mean of 8·92 mm. A gauge is set at 10·00 mm. and it is found that seven out of ninety-three ball bearings fail to pass through the gauge (i.e. have a diameter greater than 10·00 mm.). Estimate the standard deviation of the diameter of the ball-bearings?

12.8 An experiment is carried out on the crossing of two kinds of sweet pea. The varieties may be distinguished by the form of their seeds, one being round and the other wrinkled. Seed pods from the crossings are collected, and from thirty-five pods each containing eight seeds the following numbers of smooth seeds are counted:

1, 2, 1, 0, 3, 2, 2, 1, 0, 1, 2, 2, 3, 1, 4, 0, 2, 2,

3, 2, 4, 2, 3, 1, 4, 2, 3, 1, 4, 3, 2, 1, 3, 2, 4.

Use this information to test whether the proportion of smooth seeds in the population is equal to $\frac{1}{4}$.

12.9 A penny is tossed 80 times and on 45 occasions gives heads. Thus a proportion 0·5625 of tossings gives heads, whereas for a true coin this proportion should be 0·5. Test whether or not the coin could be considered to be an unbiased penny.

12.10 A die is tossed 720 times and on 143 of these tossings a six is obtained, giving a proportion 0·1986 of sixes, instead of the 0·1667 expected with an unbiased die. Carry out a test to see whether the die is unbiased or not.

12.11 It is desired to estimate the proportion of fuses unable to withstand a certain current. Four independent random samples drawn from a large number of fuses give the following results:

First sample	37 fuses drawn, 7 blow with current
Second sample	41 fuses drawn, 5 blow with current
Third sample	23 fuses drawn, 4 blow with current
Fourth sample	32 fuses drawn, 5 blow with current

(*a*) Estimate the overall proportion of fuses that would not withstand the current applied. Give the approximate standard deviation of this estimate.

(*b*) Approximately how large should the total sample be if the standard deviation as found in (*a*) is to be less than 0·01?

(*c*) Can you think of a situation in which it would be desirable to know the sample size as in (*b*)?

12.12 To make up sixty pupils for a special training course a sample of thirty pupils is selected from school *A* with a further sample of thirty pupils from school *B*. At the end of the course there is an examination which twenty-four of the pupils from school *A* pass, but only seventeen of those from school *B*. Does this indicate a real difference in the pass rate for pupils from the two schools?

12.13 A large number of patients who have not colds are available at the outset of an experiment. Two groups of sixty are selected and those in the first group are given a cold preventative. After three months it is found that whereas fifteen out of the sixty treated have had a cold, twenty of those untreated have had a cold. Do these figures demonstrate that the preventative gives any immunity from colds?

12.14 Four sets, each consisting of two dice, are thrown. A success is defined as meaning that the pips on a pair of dice add up to five. In 531 repetitions of the experiment the numbers of successes among the four sets were distributed as follows:

No. of successes	0	1	2	3	4	Total
Frequency	326	171	31	2	1	531

By calculating theoretical frequencies for the above table and comparing them with the frequencies that have been observed, test whether the dice could be considered unbiased or not.

12.15 *A* asserts that the probability of throwing a total of nine with two ordinary dice is 1/11 since there are eleven possible outcomes, namely totals of 2, 3, ..., 12 and only one is favourable. *B* asserts that the probability is 1/9 since there are 6^2 or 36 possible outcomes with two dice and four, namely 6:3, 5:4, 3:6, 4:5 give favourable outcomes.

If the statements of *A* and *B* are to be verified by tossing a pair of dice *n* times how large do you consider *n* should be, assuming that the 5 per cent significance level is to be used?

12.16 An investigation into the performance of two machines in a factory manufacturing large numbers of the same product gives the following results:

	No. of articles examined	No. of articles defective
Machine *A*	750	42
Machine *B*	900	36

Apply a statistical test in order to find out whether there is any significant difference in the performance of the two machines as measured by the number of defective articles produced. What action would you recommend if the firm concerned was considering replacing *A*, which is old, by another machine of type *B* which is, however, quite costly to install?

12.17 Ten seeds are selected from a large pile and placed on damp blotting paper. The number of seeds that germinate is noted. The whole procedure is repeated 75 times and the following results obtained:

No. of seeds germinating (k)	0	1	2	3	4	5	Over 5	Total
No. of trials with k seeds germinating	5	16	20	18	10	6	—	75

By fitting an appropriate binomial distribution see whether it is reasonable to assume that each seed has the same independent chance of germinating.

12.18 The following are the lengths (in thirty-seconds of an inch) of 12 random samples of Egyptian cottons taken from a large consignment. Test the hypothesis that the mean length of the consignment is 46 against the alternative that the mean length is more than 46.

48, 46, 49, 46, 52, 45, 43, 47, 47, 46, 47, 50.

12.19 The table below gives the results of experiments on 21 samples of iron ore using a standard dichromate titrimetric method and a new spectrophotometric method for the determination of the percentage iron ore content of the 21 samples. Test the hypothesis that the mean difference between the two methods is zero, against the alternative that it is not zero.

Sample	Standard method	New method	Sample	Standard method	New method
1	28·22	28·27	12	51·52	51·52
2	33·95	33·99	13	49·59	49·52
3	38·25	38·20	14	52·20	52·19
4	42·52	42·42	15	54·04	53·99
5	37·62	37·64	16	56·00	56·04
6	36·84	36·85	17	57·62	57·65
7	36·12	36·21	18	34·30	34·39
8	35·11	35·20	19	41·73	41·78
9	34·45	34·40	20	44·44	44·44
10	52·83	52·86	21	46·48	46·47
11	57·90	57·88			

13

SAMPLING TECHNIQUES

13.1 Some of the basic ideas of sampling were introduced in chapter 8, but these ideas can now be taken a few stages further and linked in with a wider range of practical problems. The traditional method of acquiring knowledge about an *aggregate of individuals* (this aggregate is also sometimes referred to as a *population* or *universe*) is to enumerate them all. This is especially common in the economic field, for example, censuses of population, or of imports, or of production, or of distribution. These enumerations depend upon a complete count of all the individual human beings concerned, or the goods passing through Liverpool docks, etc. Until relatively recently governments have been entirely census-minded. The major statistics of agriculture, production, distribution, of the labour force and of unemployment have all been based on the census approach. Even for prices, where an exhaustive enumeration of each transaction is impossible to carry out, the tendency has always been to include as much as possible and to spread the cover of an index number over as many commodities as could conveniently be included. The index of retail prices that is published monthly provides an example of this tendency.

But censuses are expensive, even in a small country like our own. A population census is not normally attempted more than once every ten years. What is just as bad, a census may defeat its own object by resulting in such a mass of information that the collectors are snowed under. It takes several years to analyse and publish the results of a population census, even with all the computational aids of modern electronics. The summary tables for the 1961 census of Great Britain were not published until 1966. The census-taker is always struggling against the problem of 'too much, too late'. This difficulty has, in fact, been felt so keenly that preliminary figures, based on a sample of one-in-a-hundred of the census forms completed, have in the past been issued by the census departments; and small as is the one per cent proportion, there can be no doubt that these figures are accurate enough for many purposes.

Looking further afield the problem of the census-taker becomes much vaster. The problem of acquiring comprehensive demographic (or economic) information in a country like India is so complex that a census may be a practical impossibility. Such countries are compelled to adopt sampling methods, not merely to save time or money, but as a sheer necessity. It is for this reason that in the so-called under-developed countries sampling methods are being intensively developed at the present time. UNO has found it desirable to set up a special sub-commission on sampling to aid this development.

Sampling techniques are not, however, confined to the eliciting of social or economic facts. The same sort of considerations which force them to the attention of the social statistician apply *mutatis mutandis* in scientific research, in industry and business. A manufacturer of component parts may not have the time or resources to test every item which he produces; there may even be theoretical reasons why he could not test every item if he wished. For example, if he is manufacturing shells, the only satisfactory way of testing a shell is to fire it! Again, a manufacturer of a new kind of fertiliser cannot apply it to every type of plant on every soil under every kind of climate; he has to test it under a sample of conditions.

13.2 The drawing of correct inferences from a sample about the aggregate from which it emanated boils down to one very simple problem. Something is to be asserted about a 'population'. Deliberately only a part of it is examined, usually quite a small proportion. How is it possible to ensure that a gigantic mistake is not made by neglecting the unexamined members? Should one be for ever looking over one's shoulder to ensure that nothing has been missed? Is this a sensible chance to take or is it just gambling on the possibility that the sample is representative?

There is no satisfactory philosophical answer to these questions. But it might as well be pointed out at this stage that even people who express the gravest misgivings about sampling during a philosophical discussion are quite content to rely on sampling in their ordinary daily lives. The milk you drank for breakfast may have been tested—on a sampling basis. The newspaper you read may advocate a policy based on a sample poll of the adult public. The material of the suit you are wearing was probably chosen by looking at a sample of material, perhaps of only a few square inches. And so one could go through the day, pointing out how much the quality of the goods used and the facts employed in reaching decisions are based on some sampling procedure or other. Philosophy or not, the thing seems to work in practice.

But only up to a point. One could equally easily go through the day examining the cases where the wrong kind of sample could mislead. Every schoolboy knows enough not to judge a tray of fruit in a greengrocer's shop by the specimens on top. Every reader of a newspaper must doubt whether the correspondents whose letters are chosen for print are really representative of opinion in general. No doctor would regard his patients as a fair sample of the population on any matter concerned with health or related to it.

The sampling problem in essence is relatively simple to describe. In the first place, conditions must be laid down under which the sample is a good one; that is to say, is reliable. Having done that, it is required to know how far it can be relied upon; or, to put the matter in a way which more frequently arises in practice, how big should the sample be to have sufficient confidence of being able to reach the right decision? Armed with criteria which will decide such points, the next stage of sample design can be studied so as to provide the right kind of sample at minimum cost in money, time, staff or effort.

13.3 The theory of statistical sampling rests upon the assumption that the selection of the sample units has been carried out in a random manner. Random sampling implies that the chance of any one member of the parent population being included in the sample should be the same as for any other population member; by extension, it follows that the chance of any particularly constituted sample appearing should be the same as for any other.

Long and painful experience has shown that haphazard choice is not enough to secure an unbiased sample. It is obvious that deliberate selection may bias a sample—nobody would form a view about personal expenditure on tobacco by asking men alone, or by asking men and women in a particular age-group alone. It is not so obvious that if an observer picks out individuals haphazardly he may, quite unconsciously perhaps, bias the sample.

For example, a clerk selecting a sample of invoices for detailed checking by sorting through a pile and taking one every now and then may be thought of as choosing randomly. Equally, a market research interviewer selecting women shoppers to find their attitude to brand X by stopping one and then another as they pass along a busy shopping precinct, must certainly seem to be practising random methods. In both cases, however, there is a strong possibility that bias is present in the selection of the sample. The clerk's eye may well be drawn to those invoices with elaborate printed headings, or he may pass over invoices containing many items. The market researcher may tend to ask his questions of young attractive women rather than of older housewives, or he may stop women who have packets of brand X prominently displayed in their shopping bags, rather than those with packets exhibiting the competitor's label.

Once such effects are pointed out, of course, they are easy to appreciate. Biases of such a naïve type are rarely met with nowadays except when amateurs attempt a survey. But there are more subtle effects of a similar kind to which any survey is liable unless it submits itself to a most rigorous discipline in the choice of the sample members. An interviewer who is given a list of houses to visit, for example, has to be cautioned against visiting the one next door to an assigned address if there is no reply to his knock at that address. One can see the argument from his point of view: if the houses were chosen haphazardly, the choice might just as well have lighted on the house next door as on this one, so why not take it? But obviously serious bias may appear if the quality 'being-at-home' is in any way related to the quantity under investigation in the survey, for example in investigating the amount of milk drunk per family.

In fact there is only one way to remove all possibility of bias due to bad selection; that way is to choose the sample so that every member of the population has an equal chance (or, at least, a known chance) of being included in it. This is known as random, or probability sampling. It removes from the interviewer, or whomever controls the sampled members, any element of personal choice and makes him work to a fixed set of rules. It is in the setting up of these rules, adapted to the particular requirements of the inquiry, that the primary part of the sampling design consists.

13.4 Where it is possible to identify and give every population member a number, lottery methods or tables of random numbers can be used to assist in selection of the sample. Suppose that a sample of 400 members of a shareholders' register were required. A list of all the shareholders would be needed (called the sampling frame) and a serial number would be placed against every name on the list. If lottery sampling were being used, correspondingly numbered slips of paper would be placed in a large drum and well mixed. Four hundred of these slips would be selected and the shareholders corresponding to the numbers selected would constitute the sample.

When tables of random numbers are used, the technique is somewhat faster. The tables are constructed on mathematical principles so that each digit has the same chance of selection. A table can be constructed with two ordinary dice, one red and one black, as follows. The two dice are thrown simultaneously. If the black die turns up 1, 2 or 3, then the red die gives a random digit from 0 to 4 as shown in table 13.1. A red 6 is ignored and a re-throw of the dice takes place. Similarly, if the black die is 4, 5 or 6, then the value of the red die indicates the random digit between 5 and 9 (again a red 6 is ignored and a re-throw follows).

Table 13.1. *Construction of random digit*

Black	1 or 2 or 3					4 or 5 or 6				
Red	1	2	3	4	5	1	2	3	4	5
Random digit	0	1	2	3	4	5	6	7	8	9

For example, if a four-digit random number were required the dice will be thrown four or more times. Suppose the results were:

Throw	Black	Red	Random digit
1	3	5	4
2	5	4	8
3	1	1	0
4	2	6 (re-throw)	
		2	1

The corresponding random number is 4,801. Table of random digits (not necessarily devised by the procedure just described) are available and table 13.2 gives an extract from such a table. The uses of the table will be discussed in more detail in chapter 14.

13.5 The primary importance in probability sampling of avoiding bias has been stressed. The method has one other fundamental

Table 13.2. *Extract from table of random numbers*

61	22	64	33	71	83	30	50	23	05	06	00	50	03	97
01	61	25	35	87	89	53	35	34	65	41	51	04	63	35
02	09	39	82	74	40	69	03	04	95	33	67	53	43	10
98	00	23	11	92	99	82	84	35	07	52	83	94	20	33
03	99	12	29	93	71	58	33	13	22	87	76	09	35	00
29	56	29	84	17	27	66	35	46	42	61	45	06	13	72
94	03	39	52	49	30	76	86	53	55	13	79	21	23	00
48	03	09	97	42	08	68	39	00	29	68	59	10	07	46
04	40	98	54	35	88	97	92	79	91	28	50	02	17	54
45	06	61	14	31	38	00	14	41	90	60	25	71	32	43
12	13	18	26	21	12	96	88	01	15	31	70	42	92	74
00	84	71	68	07	95	71	88	31	72	06	99	75	04	74
18	36	80	94	87	41	42	16	87	40	62	36	87	84	42
16	64	33	06	19	31	91	55	66	12	82	29	39	21	91
83	81	30	51	43	63	89	08	98	66	42	09	01	04	79
21	27	31	57	40	60	75	96	16	86	82	80	32	00	00
31	15	57	80	78	79	27	34	25	63	87	36	08	89	28
86	48	99	72	28	95	25	84	30	50	32	33	78	77	64
35	01	74	20	41	84	47	70	04	45	72	22	03	72	80
15	87	19	42	15	44	52	02	66	22	64	86	52	02	85

advantage of equal importance. With this method, and this method alone, the precision of the results can be assessed in objective terms.

Again the point is a very simple one. Suppose the shareholders' sample of 400 surnames had 30 of them beginning with the letter M. The sample is a random one and, noting that 7·5 per cent have the initial M, it seems reasonable to estimate that this is the proportion with the initial letter M for the (large) population of shareholders. But might the true proportion not be 7 or 8 per cent; or even 6 or 9 per cent? The theory developed in chapter 12 shows that, if p is the proportion of individuals with the initial M, n is the sample size and x is the number in the sample with letter M, then the standard deviation of x/n is:

$$\sqrt{\left(\frac{p(1-p)}{n}\right)}.$$

If the sample estimate, 0·075, is put in as being an approximate but appropriate estimate for p, this gives 0·013. Hence the normal distribution would suggest that, with 90 per cent certainty, the possible range of the true but unknown value of p is:

$$0{\cdot}075 \pm 1{\cdot}645 \times 0{\cdot}013 \quad \text{or} \quad 0{\cdot}054 \text{ to } 0{\cdot}096$$

If the sample were 4,000 instead of 400 the band would be somewhat narrower, i.e. 90 per cent certainty that the true proportion fell in the range 0·068 to 0·082. This band's width is in the ratio of $1/\sqrt{10}$ to that of the previous band; a result that must be expected from the form of the expression used above for the standard deviation.

What width of band is acceptable in real life is a matter for discussion in each particular instance. A cautious investigator may want a narrow band; but if so, he has either to accept a lower degree of confidence or a much larger sample size. Thus, although it is not possible to make a statement from sample to population with the certainty associated with deductive logic, it is possible to make a precise statement in probability and such statements are for most practical purposes sufficient.

13.6 Considerations of time and expense may often make the task of conducting a simple random sample prohibitive. A simple random sample of 1,000 electors in the United Kingdom could involve visits not only to different parts of many towns, but also to outlying and remote regions far scattered from each other.

Again there is the worry when conducting random sampling that an 'unrepresentative' sample will result. For example, if one took a random sample of 100 students from the 5,000 registered at a particular university, it could just conceivably happen that all 100 were first-year arts students. The method of drawing the sample may conceptually have been quite correct, yet the sample could not be thought of as representative of all students in the university. Where there are well-defined categories within the population to be sampled, such as year of course, type of course, etc., the possibility of an unrepresentative sample can be avoided by taking a pre-determined number of sample units from each section (or stratum). The accuracy of the result, as measured by the standard error, will then be improved.

Suppose that the area of land under wheat in the United Kingdom is to be estimated, by sampling from the 'population' of farms. If farms were picked at random there would be a lot of small ones and only a few large ones; but the estimate of area would be very sensitive to the number of large farms, and the occasional very large farm should not be missed. To overcome this, group the farms into strata by size; there is then nothing to prevent a different sampling fraction being adopted for each stratum, selecting say one in 100 for farms from 50 to 100 acres and every one of those above 1,000 acres. By balancing the stratification and the randomness in this kind of way the best of both worlds can be achieved, without incurring the risk of bias.

13.7 Another common modification introduced to simple random sampling is that of multi-stage sampling. This, as will be shown, helps to ensure that the field-work of a survey is carried out in a limited number of areas or institutions with a consequential saving of time and money. For example, if a sample of individuals in England were required one might first of all choose a number of counties at random and then, within each county, choose a defined number of polling districts at random and then, within each selected polling district, choose a number of names from the appropriate electoral register. This, properly done, can be made to yield a sample such that every member of the population has the same chance of being chosen. It is true that certain counties will not appear at all; but they had a chance of doing so and in the long run will do so when the sampling is repeated from occasion to occasion.

The inherent disadvantage of multi-stage sampling is that the standard error will be greater than for simple random sampling with the same sample size. The cause of this is not hard to locate. When human beings or inanimate objects are considered in groups, it is almost invariably the case that the groups will be made up of people exhibiting the same environment, background and so forth. If roads were used at the first stage of a sample, it is easy to appreciate that each road will tend to contain houses of a similar nature with regard to size, price, etc. It is rare to find one road with the full range of housing types within it. Because of this it is quite possible that a highly unrepresentative sample may be drawn, and it is therefore better to take small samples from a large number of groups than large samples from a small number of groups. It also follows that the individuals within the groups should be as heterogeneous as possible, which is the opposite requirement from stratified sampling where sampling units within strata should be as homogeneous as possible.

13.8 All the sampling designs so far discussed rest upon random or quasi-random sampling, but in commercial surveys this is often rejected in favour of the following method of *quota* sampling. Here the selection of the sample units from the frame is no longer dictated by chance; indeed, a sampling frame is not used at all, and the choice of the actual sample units to be studied is left to the direction of the interviewer or investigator concerned. He or she is, however, restricted by 'quota' controls. One interviewer may be told to interview seven married men, between the ages of 35 and 45, who are manual workers and live in town Y. Obviously these characteristics apply to many men, but the selection of the 7 men will be left to the discretion of the interviewer.

The advantages of quota sampling are self-evident. There is no need to have available, or to construct, a frame; furthermore, because a frame is not being used, it is possible to implement a greater degree of stratification in setting up the controls. Age and sex are not given in electoral lists, but can be built into quota controls. Falsification of returns is, however, a greater danger than with the normal methods of sampling.

The superficial resemblance of quotas to the strata of probability sampling must not obscure two basic differences: (i) selection within the quotas is not randomised selection from frames within strata, but is directed by the judgment of interviewers; (ii) it is no longer theoretically possible to evaluate a sampling error. Of course the lower costs per unit are often obtained at the price of higher variability, particularly as the lower the degree of quota control the lower the cost per interview. Interviewers' experiences abound with tales such as 'At our local park I can always get over a dozen good interviews in an afternoon, from a sample of mothers sunning their babies'.

In spite of their many faults, quota samples do produce some good results. The bold predictions of elections have frequently come close to the actual election results. They often represent considerable improvement over other forms of judgment sampling such as journalists' haphazard encounters, a 'typical' town, a poll of volunteer correspondents, or of a magazine's subscribers. Quota controls may help to spread the sample in ways that might not otherwise happen. A quota sample from young people in general is more likely to represent the attitude of the nation's young people than a probability sample from amongst a university's students.

Each of the following two sections discusses a recent sample survey in order to illustrate some of the points made in this and earlier sections.

13.9 The first example is taken from the Robbins Committee on Higher Education which reported in 1963. The committee wished to carry out a survey of all students taking advanced courses at institutions of further education. It was decided to aim at a final sample of 5,000 students (1,500 on continuous full-time courses, 1,500 on sandwich courses, 1,000 on part-time day courses and 1,000 on evening courses). The sample design involved two stages of selection. The first stage units were colleges or, in some cases, groups of colleges (where colleges were very small) which had students taking advanced courses, and the second stage units were individuals. The method of selection at each stage was random.

At the first sample stage 91 colleges were selected in the following manner. The colleges (and groups of colleges), listed by the Ministry of Education and Scottish Education Office, were divided into five strata by type of institution. These are shown in stage I of table 13.3. At the second stage, individuals, stratified by method of study (continuous full-time, sandwich, part-time day and evening) were chosen from the colleges selected at the first stage. Lists of students were provided by each selected college. Variable sampling fractions were used to give, within each college, an overall sampling fraction of 1:12 for students on continuous full-time courses; 1:6·7 for sandwich course students; 1:45 for part-time day students; and 1:38 for evening students. The ratios were chosen so as to yield the numbers required and were based on 1960/61 student numbers. The resulting sample sizes, shown in stage II of table 13.3 produced a total of 5,961 students in the sample owing to a rise in student numbers over 1960/61.

The overall response rate to the circulated questionnaire and interview was 91 per cent, varying between 86 and 94 per cent for the four categories of student. The selected colleges were approached in January 1962 and the first students were interviewed the following month. It had been hoped to complete the field-work by April, but some colleges were late in supplying lists of students and field-work continued until the end of May. At this point, information was not available for 7 of the selected colleges and it was decided to omit these rather than risk further delay.

One difficulty arose in that some students reported a different

Table 13.3

Stage	Unit	Stratification	Number of units selected	Method of selection
I	Colleges or group of colleges	Type of college:		
		Colleges of Advanced Technology	9	All
		Regional colleges and regional colleges of art	16	1 in 2
		Other colleges in England and Wales stratified into: (i) Counties (ii) County Boroughs	58	Proportional to number of advanced students in 1960/61
		Central Institutions (Scotland)	6	1 in 2
		Further Education Centres (Scotland)	2	Proportional to number of advanced students in 1960/61
		Total	91	
II	Individuals	Method of study:		Variable sampling fraction giving an *overall* sampling fraction of
		Continuous full-time	1,930	1:12
		Sandwich	1,681	1:6·7
		Part-time day	1,006	1:45
		Evening	1,344	1:38
		Total	5,961	

Note: The selection at each stage was random.

method of study from that on the college register. Any re-allocation of students according to the method of study they themselves reported would have involved re-weighting the sample, and this would have caused an amount of work disproportionate to the gain in accuracy. Accordingly these returns (20 of them) were excluded from the analysis.

13.10 A second illustration is drawn from National Opinion Polls Ltd. who conduct regular political surveys, the results being given with appropriate commentary in the *Daily Mail*. Again a two-stage procedure was used, the first step being to select 100 constituencies out of the 618 in Great Britain. This is done by listing the constituencies according to the Registrar General's ten standard regions and Scotland. Borough constituencies were placed separately from County constituencies. Within the resulting 22 categories, constituencies were listed according to the ratio of Conservative to Labour votes cast at the General Election of 1964. The sample was drawn on a systematic probability basis so that each constituency had a chance of selection proportionate to the number of electors it contained. This was done by taking a randomly determined starting point in each of the 22 lists and then working through the list at the average number of electors between samples, i.e. $\frac{618}{100} \times$ average number of electors per constituency. This method ensures that the correct proportion of Borough and County constituencies are drawn within each region and that within groups of Borough or County constituencies in any one region, a wide cross-section of constituencies is selected—varying from the safely held seats to the more marginal ones.

Although the same 100 constituencies are used in each survey, the sample of 30 electors taken from each is different every time. The method of selecting these from the electoral register is a modification of systematic sampling and varies between County and Borough constituencies. In the County constituencies a random number between 1 and the total electorate is taken and the person corresponding to this number on the register together with every tenth person following (up to a total of 30 in all) is included. In the Borough constituencies the same procedure is followed, but only 15 sample units are selected. The second 15 are chosen by first adding to or subtracting from the random number half the total electorate; this number determines the 16th person to be included, the remaining 14 being at ten-person intervals thereafter.

No substitutes are allowed when sample units are out or away from home. As a result of several re-calls, an 85 per cent response rate is obtained from those people still living at the address given in the register. Supervisors are employed to check the work of the field staff.

13.11 Finally, some rather more brief notes on various sampling tasks are given as further illustrations.

(*a*) A sample of accounts was required through which to study the revenue that accrued to the carriers of freight-shipments of various commodities between different sections of the country

(U.S.A.). When a shipment traverses two or more railways, the railways that participate in the shipment divide the total revenue for the shipment amongst themselves according to a complex formula. The reason for making this study was that certain railways were threatened in a lawsuit with a change in the formula that would have brought a reduction in the share of the revenue that they derived from those shipments that traversed some portion of their tracks. The aim of the study was to furnish evidence by which to predict, on the basis of past records, the future effects of the proposed reduction.

(*b*) A manufacturing firm has in one plant an inventory of materials in progress valued at about £7m. The firm can decrease taxes and obtain more meaningful information for management by estimating the change in value of this inventory over the course of the year. The sampling problem is to design a sample of the material held that will provide the desired information with the precision required.

(*c*) A sample was required by the Government to assess food consumption and expenditure. It was necessary to cover households of different family composition and social class, and to take into account their distribution by region and type of area. This was done by a three-stage scheme: (i) Selection of 50 parliamentary constituencies; (ii) Selection of a number of polling districts within these constituencies; (iii) Selection of a number of households within each of the polling districts chosen at the second stage.

(*d*) Study of tax accounts in certain ledgers. There were 23 large volumes, each containing around 500 pages. There were about 27 lines to a page, names appearing on many of the lines. These names constituted the catalogue of listings. Some names showed a money entry, indicating the tax unpaid at the date of posting. The study aimed to estimate the number of money entries by size so as to provide a basis for deciding whether to try to collect any of the amounts due and, if so, which sizes of outstanding amounts, aiming to maximise revenue obtained thereby for given effort.

(*e*) The G.P.O. carried out a survey to establish what part telephone directories played in finding the appropriate telephone number in a certain region. They found the part played was limited and estimated that if new directories were not issued for a year they would not be greatly missed and that an increase of not more than 1·3 per cent in the use of the inquiries service would result.

(*f*) A large company is contemplating a group life assurance scheme and wants to know, amongst other things, the average amount of life assurance carried by its employees. A simple random sample could easily be selected using, say, the company's payroll. It is likely, however, that the average amount of assurance is quite different for the various categories of employees and this fact could be exploited to obtain a more efficient sampling design (i.e. greater information for given effort or the same level of information for less effort) by use of stratification techniques.

EXERCISES

13.1 A Broadcasting Company has used the following methods to find out how the programmes are appreciated:

(i) Inviting casual passers-by to speak into a portable microphone at a busy street corner and express their likes and dislikes.

(ii) Scrutiny of letters in the Press and letters received by the Company from listeners.

(iii) Door-to-door inquiries in selected areas by a trained staff of interviewers.

(iv) Employment of a small body of paid listeners who are required to report on reception, interference and other technical matters as well as on the programmes themselves.

Criticise briefly each of these methods.

13.2 (i) What features in the design of a statistical inquiry tend to reduce variations due to random sampling?

(ii) Explain how you would use a table of random numbers to select a sample of n from a population of 7,500.

(iii) Indicate what bias, if any, is likely to arise in sampling in the following circumstances:

(*a*) a crop survey in which the local agricultural officials have been asked to select a representative sample of fields;

(*b*) a survey of a random sample of houses by personal interview in which the next house in numerical order in the street is taken when there is no reply at the selected house;

(*c*) a postal questionnaire.

(iv) In what circumstances would a small amount of bias be acceptable?

13.3 A market research survey is to be made into the attitude of adults in a certain region towards a particular product. A questionnaire has been drafted for use by 50 interviewers. Each interviewer has been allotted an area and is instructed to obtain replies to the questionnaire during a given week from 100 persons chosen at random in this area. The 50 areas give representative coverage of the whole of the region.

(i) Draft a concise letter to the organisers of the survey explaining why the instructions given are unlikely to produce a satisfactory sample. Describe in your letter, with reasons, the method of sampling which you consider would be the most suitable in order to obtain a representative sample on the basis of: (*a*) house-to-house inquiries; (*b*) street inquiries.

(ii) Discuss the factors that must be considered in deciding upon the number of interviews in a survey of this type.

13.4 Records of the children attending the schools in a certain area give, for each child, name, address, date of birth and name of school. These records, which may readily be sorted into any required order, are to be used to draw a sample of schoolchildren. Any additional information would be obtained by inquiry in respect of each child chosen.

What methods of sampling would you recommend for the purpose of estimating: (*a*) the proportion of blue-eyed children, (*b*) the proportion of children living at least two miles from school; (*c*) the ratio of actual attendances to possible attendances in a given period?

Give reasons for each recommendation and indicate briefly how the recommended methods would be applied.

13.5 (i) Describe the qualities required in a sampling frame.

(ii) Discuss the use of the Register of Electors as a sampling frame for a survey designed to measure the prevalence of cigarette smoking in Great Britain.

13.6 Outline possible sampling methods which might be devised to help in investigating the following problems:

(i) The management of a gas works wish to analyse the coal supplied from various collieries (*a*) by railway wagons, and (*b*) by barges.

(ii) A manufacturing chemist requires to know the purity of the common salt supplied to him in 5 cwt. sacks.

(iii) An excise officer has to measure the alcohol content in whisky produced by a certain distillery.

13.7 A life assurance company wishes to know how many policies on the average are effected on each separate life assured with the company. The available records include (*a*) a list of policies in numerical order with particulars of the persons on whose life each policy was issued, and (*b*) a card index in alphabetical order of surnames of the various lives with particulars on each card of all policies issued on that life.

How would you proceed to investigate the problem by sampling and what precautions would you take to avoid bias?

14

SIMULATION

14.1 The usefulness of probability distributions was well established by such early mathematicians as Laplace (1749–1827) and Gauss (1777–1855). The idea that frequency distributions could be explained as a practical consequence of the laws of probability applied to everyday matters, seized the imagination of the pioneers of mathematical statistics. Since a probability distribution is by its nature, in most instances, composed of an infinite number of items, and frequency distributions by their nature are composed of a finite number of items, these latter had to be thought of as samples from an underlying theoretical probability distribution. The problem that then arose was how to describe a probability distribution given only a sample from it. The mathematical difficulties of this seemed immense and such steps as were taken needed experimental verification to give the early workers confidence. Thus was born the sampling experiment. A close approximation to a probability distribution was created, samples were taken, combined and transformed in suitable ways and the resulting frequency chart of sampled values compared with the predictions of theory. Although mathematical techniques have developed to levels of sophistication that would astonish earlier workers, the value of sampling experiments in mathematical statistics still remains.

There are two main types of distribution from which samples are required. The first is where the statistical variable takes a continuous form, giving rise to a continuous probability density function; the second is where the statistical variable can take a discrete number of values. The normal distribution is typical of the first class and the binomial distribution (the number of successes in a fixed number of similar independent trials) is an example of the latter. However, for practical purposes it is necessary to work with approximations to continuous variables and to have to be satisfied with samples from a grouped distribution represented by a histogram. The problem then reduces to selecting each of these groups with the appropriate frequency and consequently there is no distinction in practice between the two types of distribution.

The finding of a sampling distribution by experimentation in this manner is referred to as *simulation*. It is a process of selection such that the results of the repetition of this process over a period will give rise to a frequency distribution of sampled values that matches the frequency distribution required. It is possible for there to be quite intricate probability rules relating successive sample values, but the most important cases which will be considered in detail here are where the samples are independent of each other, in the sense that the probability distribution for any given item from the sample is independent of the preceding sample values.

14.2 A standard technique of random selection is the system commonly referred to as the Monte Carlo technique, called after the gambling casinos rather than the car rally. To illustrate the principles, a simple example will be described. Consider a new product which contains two independent and distinct parts, each of which will eventually fail. These parts might be a condenser and a vacuum tube. From past tests and records, the probability of failure of each item in terms of its time in use has been estimated, i.e. the life curve of each item is available. What is wanted is the life curve of the product which contains one of each of those elements—assuming the product fails when the first component fails. Denote the original life distributions of the components by the symbols f and g respectively, whilst the symbol h denotes that of the combined derived distribution. Now in some cases h can be derived by mathematical analysis. This will occur when f and g can be represented by certain simple mathematical functions. But in other instances it is not possible or practical to evaluate the derived function h in this way. In such instances the Monte Carlo technique may be the best method available. Assume that the frequency of incidence of time to failure of the two components are as shown in fig. 14.1. In the assembly of the product, one item from each class of component will be selected at random.

To simulate the formulation of assemblies, items for each component must be selected in such a way that each has an equal chance of being selected. Since there are clearly more items of the first component, f, with life spans in some intervals (say the interval from 95 to 105) than in other intervals (say the interval from 75 to 85), the procedure must be such that the chance of selecting an item in any chosen interval on the time scale is equal to the pro-

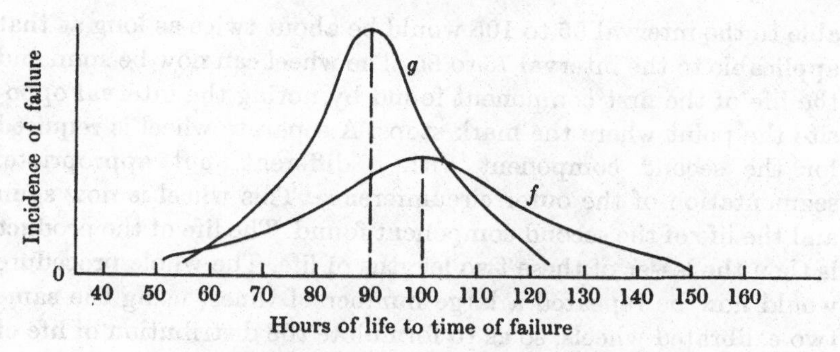

Fig. 14.1. Distribution of component lives

portion of items falling in that interval. Now if a simple random sample were taken from the set of possible time values laid out along the horizontal axis, this would lead to giving the same chances to the drawing of an item with a life span between 95 and 105 as to one between 75 and 85. This is clearly both an inflexible and an incorrect procedure in this instance. Hence a technique must be devised for sampling from frequency distributions, taking into account the relative frequencies of the different intervals.

14.3 One time-honoured way of achieving this selection is by drawing discs from a drum. The discs would have numbers on them representing the possible values of the variable (e.g. time) concerned. To achieve the necessary weighting of the values of this variable, the number of discs given to any particular value would need to be proportional to the frequency with which such a value occurs in the frequency distribution in question. In theory a large number of discs would be needed, to ensure a sufficiently good representation of the distribution, and the discs must be well shuffled before each drawing to ensure that each disc stands an equal chance of being selected. To obtain a series of values for the variable concerned, a series of discs would be drawn, each drawn disc being replaced and the drum reshuffled before the next disc is drawn.

An alternative method of achieving the same result would be to take a roulette wheel, with a mark or arrow on one point of the inner spinning circumference, and to segment the outer numbered circumference in proportion to the chances required for different lengths of life. For example, the segment of circumference applicable to the interval 95 to 105 would be about twice as long as that applicable to the interval 75 to 85. The wheel can now be spun and the life of the first component found by noting the interval opposite the point where the mark stops. A separate wheel is required for the second component with a different, but appropriate, segmentation of the outer circumference. This wheel is now spun and the life of the second component found. The life of the product is then the lesser of these two lengths of life. The whole procedure would now be repeated a large number of times, using the same two calibrated wheels, so as to formulate the distribution of life of the product. The resulting distribution for h is of the form shown in fig. 14.2. To obtain a distribution as smooth as that shown

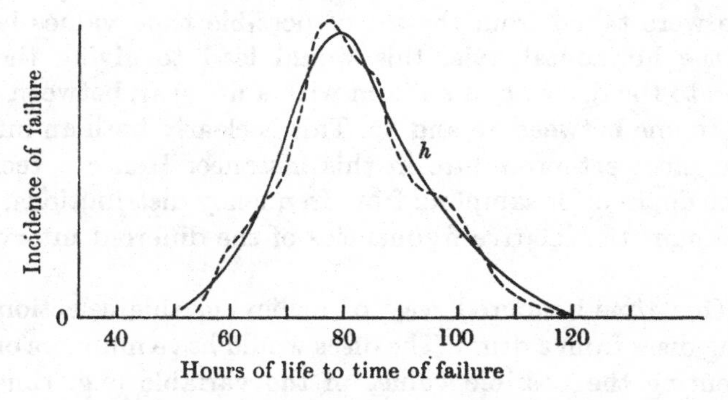

Fig. 14.2. Distribution of combined product lives

would require a large number of simulations, and the jagged dotted distribution shows the form that it might have reached at a somewhat earlier stage. Note that these techniques can be applied whatever the shape of the original distributions of component lives. The distributions do not have to be smooth and bell-shaped as shown, but could equally well be of an irregular form.

14.4 In practice the shuffling or wheel spinning can be avoided by visualising each disc as having a serial number, as well as the value of the variable. The random selection then turns on ensuring that the serial numbers of the discs chosen are a random sample. This can be taken a stage further and for the drum of numbered discs a table of values or random variables, each one numbered serially, can be substituted. To apply this, fig. 14.3 (which is

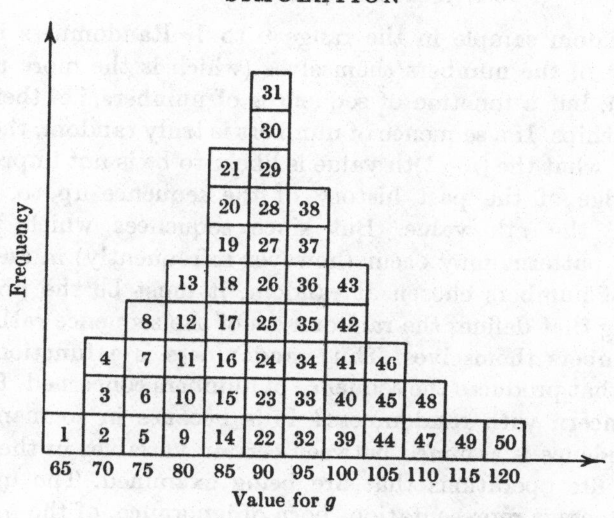

Fig. 14.3. Random sampling skeleton

an approximate representation of component g from fig. 14.1) shows the variable g being allotted serial numbers according to the frequency of each value, with the lower values of the variable being allocated the lower serial numbers. In the figure the only critical points now are those serial numbers with which a new value of the variable g is associated. This enables the so-called histogram of fig. 14.3 to be turned into the following rather simpler tabular form:

Value of g	65	70	75	80	85	90	95	100	105	110	115	120
Lowest serial number	1	2	5	9	14	22	32	39	44	47	49	50

The serial number scale could be changed so that it ran from 0 to 1 instead of 1 to 50, by dividing all numbers by 50. The problem has then been reduced, by a common mathematical device, so that sampling from any defined distribution is put in terms of the simpler and more generally applicable problem of selecting random numbers uniformly in the range 0 to 1. If a method is available for the latter problem, it can accordingly be generalised to cover sampling from any distribution.

14.5 A sequence of random numbers (or more precisely a random sequence of the digits 0 to 9) may be used to give the decimal representations (to any desired degree of precision) of the members of a random sample in the range 0 to 1. Randomness is not a function of the numbers themselves (which is the more random, 7 or 4?), but a function of sequences of numbers, i.e. their interrelationships. If a sequence of numbers is truly random, the knowledge of what the $(n+1)$th value is likely to be is not improved by knowledge of the past history of the sequence up to, and including, the nth value. But since sequences which produce marked patterns may occur (however infrequently) in an infinite series of numbers chosen at random, it must be the process of choosing that defines the randomness of the sequence rather than the numbers themselves. Thus randomness is a function of the device that produces the sequence of numbers concerned. But why this concern with randomness? It is because in so many cases independence is assumed between certain variables in the models of real life operations that are being examined. The quest for randomness is representation, born of ignorance, of the imperfect independence, and hence does not itself generally have to be perfect. Ideally at any point of the sequence each number should be equally likely to occur, irrespective of the pattern of numbers earlier in the sequence.

14.6 The principles of simulation will now be applied to the following problem. Random samples of size four are to be selected from a normal distribution with mean zero and standard deviation of one (the unit normal distribution). What is the mean, median and upper and lower quartiles of the corresponding distribution of the *range* (i.e. highest minus lowest value)? It is true that this problem can be solved by mathematics. But it will be instructive to approach it by simulation and to compare the results with those obtained by the exact mathematical approach.

To carry out the simulation it is first necessary to construct a table that will convert random numbers into the appropriate normal variable. To do this, table 9.4, of the unit normal distribution, is used and the values of x associated with a range of four-digit random numbers with band 0000 to 9999. Consider, for example, the small portion of the table which reads:

x	$F(x)$
$-2{\cdot}0$	$0{\cdot}0228$
$-1{\cdot}9$	$0{\cdot}0287$.

Table 14.1. *Conversion of unit normal distribution*

x	Random numbers	x	Random numbers	x	Random numbers
−3·39	0000	−1·15	1151	+0·95	8159
−3·15	0007	−1·05	1357	+1·05	8413
−3·05	0010	−0·95	1587	+1·15	8643
−2·95	0013	−0·85	1841	+1·25	8849
−2·85	0019	−0·75	2119	+1·35	9032
−2·75	0026	−0·65	2420	+1·45	9192
−2·65	0035	−0·55	2743	+1·55	9332
−2·55	0047	−0·45	3085	+1·65	9452
−2·45	0062	−0·35	3446	+1·75	9554
−2·35	0082	−0·25	3821	+1·85	9641
−2·25	0107	−0·15	4207	+1·95	9713
−2·15	0139	−0·05	4602	+2·05	9772
−2·05	0179	+0·05	5000	+2·15	9821
−1·95	0228	+0·15	5398	+2·25	9861
−1·85	0287	+0·25	5793	+2·35	9893
−1·75	0359	+0·35	6179	+2·45	9918
−1·65	0446	+0·45	6554	+2·55	9938
−1·55	0548	+0·55	6915	+2·65	9953
−1·45	0668	+0·65	7257	+2·75	9965
−1·35	0808	+0·75	7580	+2·85	9974
−1·25	0968	+0·85	7881	+2·95	9981
				+3·20	9987

This implies that the area up to $x = -2{\cdot}0$ is 0·0228 whilst that up to $x = -1{\cdot}9$ is 0·0287. Hence the area between $x = -2{\cdot}0$ and $x = -1{\cdot}9$ is 0·0287 − 0·0228 or 0·0059 and could be represented by the random numbers 0228 to 0286 inclusive (i.e. 59 such numbers). The corresponding value of x for all these numbers would be put at −1·95, i.e. mid-way between the two end points of the frequency group. If this procedure is repeated throughout table 9.4, the results given in table 14.1 are obtained. To save space, and to follow the notation in section 14.5, only the lowest random number corresponding to each value of x is given. The only two values of x which pose any difficulty in this classification are the two extreme values. The random numbers 0000 to 0006 inclusive should correspond to the average value of x within the range $(-\infty)$ to $(-3{\cdot}2)$, the average being a weighted average with weights corresponding to the probabilities of the various possible values of x. Table 9.4 does not enable one to perform this calculation and a more extended table has been used to estimate the appropriate value of −3·39 given for x in table 14.1 against the first group. A similar calculation has been made for the last group. Note that extreme accuracy is not really necessary as only a very few sample values will fall in these two groups. For the other groups the assumption that the average value of x is the same as the mid-point between the two boundary values of x will not produce any meaningful error in the method, since over such a small range of values of x the probability density function will be virtually uniform.

14.7 To draw a sample from the conversion table of table 14.1, the table of random numbers given in table 13.2 is used, taking groups of four digits from the first four columns and reading downwards. Thus the first such random number is 6122. From table 14.1 this corresponds to an x value of $+0{\cdot}25$. The next random number is 0161 which corresponds to an x value of $-2{\cdot}15$. Proceeding in this way the first random sample of four x values is built up as follows:

Random number	x value	Range
6122	$+0{\cdot}25$	
0161	$-2{\cdot}15$	$2{\cdot}05-(-2{\cdot}15) = 4{\cdot}20$.
0209	$-2{\cdot}05$	
9800	$+2{\cdot}05$	

The reader is left to verify that, proceeding in the same manner, the next sample of four x values has a range of $1{\cdot}20$. Notice that, in practice, once the group of four sets of random numbers is established, only the highest and the lowest of these need be translated to the corresponding x value in order to determine the range of the sample (i.e. in the illustration above only the x values corresponding to the random numbers 0161 and 9800 need to be known). The procedure can now be repeated using fresh random numbers. Only 37 samples can be obtained using table 13.2, but further sources of random numbers are available and are discussed later in the chapter.

14.8 When 100 such samples had been obtained the values of the range were grouped up into the frequency distribution shown in table 14.2. From this distribution, or better still from the 100 original values of the range that went to make up this distribution, the various parameters of the distribution can be estimated. These are tabulated in table 14.3 which gives the four parameters estimated, first, from the 100 sample values of range grouped as in

Table 14.2. *Observed distribution of range*

Range	No. of observations
0–	3
0·5–	7
1·0–	16
1·5–	24
2·0–	19
2·5–	16
3·0–	6
3·5–	3
4·0–	3
4·5–	2
5·0–	1
5·5–	0
Total	100

table 14.2; and secondly, the theoretical values obtained from a mathematical derivation of the distribution of range (not dealt with in this book). As will be seen, the results from the simulation are very close to the exact results. If more range values were simulated the agreement would very probably be even closer. The need to do extra simulation must, therefore, depend upon the degree of accuracy required from the simulation.

Table 14.3. *Comparison of constants estimated for range*

Constant	Simulation sample (section 14.8)	Mathematical calculation	Simulation samplo (section 14.9)
Mean	2·01	2·06	2·02
Median	1·99	1·97	1·99
Upper quartile	1·47	1·41	1·46
Lower quartile	2·70	2·62	2·68

14.9 Clearly one factor in the accuracy of the simulation rests in the conversion table groupings used to convert random numbers to values of the variable x. In table 14.1 there were 64 groups (or possible values of x). If the table had been a very much cruder table with only, say, a dozen groups as follows:

x	Random numbers	x	Random numbers
−3·3	0000	0·3	5000
−2·7	0013	0·9	7257
−2·1	0082	1·5	8849
−1·5	0359	2·1	9641
−0·9	1151	2·7	9918
−0·3	2843	3·3	9987

then it must be anticipated that the values of range obtained would have a greater degree of rounding error contained in them. Hence rather more simulations would be required to achieve the same level of accuracy as before in the estimation of, say, the median or any of the other parameters.

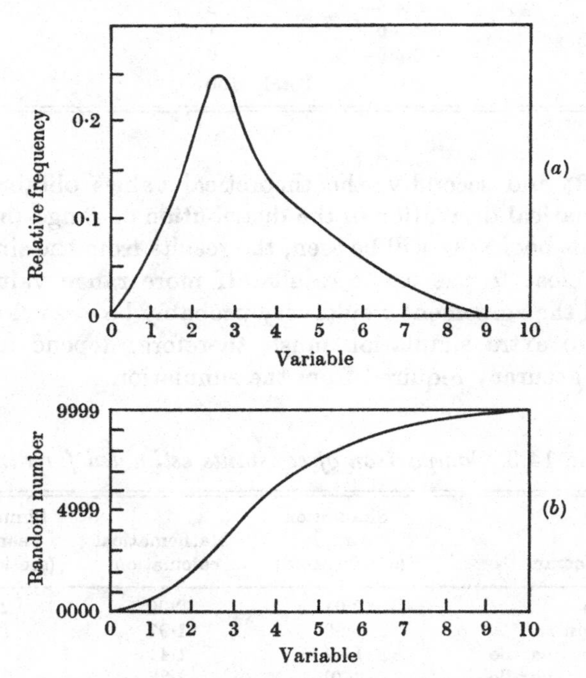

Fig. 14.4. Cumulative distribution conversion

But this doubt suggests that it may be better in some circumstances to simulate direct from the cumulative distribution if this can be done. Fig. 14.4 (*a*) shows a frequency distribution corresponding to a variable from which it is desired to draw random samples. The corresponding cumulative distribution is shown in fig. 14.4 (*b*). If the vertical scale is made to read from 0000 to 9999 in place of 0 to 1·0, this will give a distribution that is approximately uniform over the range 0 to 1. A random number of four digits is selected and located on the ordinate. Reading across the figure, the corresponding abscissa is read off. Thus the random number of 4,127 would correspond to an x value of 3·1 and the random 8,983 would correspond to an x value of 6·9. (It may be difficult to read off the values to this accuracy from the small version of the cumulative distribution reproduced here.) Viewed in this light, it can be seen that the problem of sampling from any distribution is that of transforming a random number representing the uniform random variable in the range of 0 to 1 by means of the appropriate cumulative distribution function. Using this method for the previous illustration, the results shown in the last column of table 14.3 were obtained.

14.10 Since many simulations are large-scale affairs, vast quantities of random numbers are needed, well beyond the modest table provided in chapter 13. Over the years various books or pamphlets of random sampling numbers have been produced, of which the best known are:

Random Sampling Numbers, by L. H. C. Tippett (Tracts for Computers No. XV).
Random Sampling Numbers (2nd series), by M. G. Kendall and B. Babington Smith (Tracts for Computers No. XXIV, 1946).
A Million Random Digits, The Rand Corporation (The Free Press, 1955).

The first of these has 10,400 digits; the second 100,000 digits; and the third, as its name implies, 1,000,000 digits. Frequently, however, simulation procedures are allied to using computers and for this purpose the random numbers would, on the face of it, have to be stored in the computer's memory. The quantities of such numbers required may be very large and hence exhaust the size of the computer's memory, besides being a very inefficient way of using the memory. Consequently, computer programs have been developed to produce within the computer a series of digits by deterministic methods; such digits are referred to as *pseudo-random* numbers. There are quite a few procedures that have been devised for generating such a series. One popular method is the so-called 'mid-square' procedure. This consists of selecting a four-digit number, preferably (but not necessarily) taken at random as the first of the series. This number is squared and the four digits starting at the third from the left form the next random number, and so on.

For example, if the starting number is 3,182, squaring this gives the number 10,125,124. The four middle digits 1,251 are recorded, squared, and so on. Eventually this procedure will return to the number with which it started, the length of the cycle generally falling between 10^4 and 10^6. This method of generation has some theoretical drawbacks which throw doubt on their true randomness and, as a consequence, some more complex methods of generation have been devised. Most computers have good programs available for random number generation which makes simulation of sampling experiments fairly straightforward.

14.11 In most instances where simulation is being employed to estimate the parameters of a sampling distribution, the results cannot be analytically determined. Therefore the researcher cannot determine precisely in advance how many simulation 'runs' he will need. For this reason, the application of sampling in stages (at least in two stages) is particularly well suited for simulation. When the simulation is done by hand, the variance of the estimate sought can be recomputed after each trial. When a computer is used, however, it is costly and time-consuming to divide the simulation into stages. What is frequently overlooked, on the other hand, is that it may be more costly and time-consuming to select a sample size arbitrarily and to make too many or too few observations. Even when a computer is used, several hand-runs of the simulation are usually required in order to check out the computer program. These can also be used to obtain at least a crude estimate of the variance, which can then be used to estimate the required sample size.

EXERCISES

14.1 A man travels on an underground train and then takes a bus. The distributions of times including waiting time, for the two journeys are given on p. 227.

Assume that the time taken on the train is independent of that taken on the bus. Use simulation to estimate the upper and lower quartiles of the man's overall journey time. (Carry through 40 simulations of a journey.)

Train		Bus	
Time (min.)	% frequency	Time (min.)	% frequency
26	10	9	5
28	20	10	10
30	30	11	20
32	20	12	30
34	10	13	20
36	6	14	10
38	2	15	5
40	2		

14.2 The claim arising from a group insurance policy each year may be assumed to come from the following distribution:

Claim amount (£)	Relative frequency
0–1,000	9
1,001–2,000	18
2,001–5,000	33
5,001–10,000	20
10,001–20,000	11
20,001–50,000	7
50,001–100,000	2

Use a simulation method to draw 10 random samples, each sample consisting of 5 claims. For each sample establish the largest claim and find the mean and standard deviation of these 10 largest claims.

14.3 Using the samples drawn in exercise 14.2, find the mean claim per sample and thence the mean and standard deviation of these mean claims. What is the theoretical mean and standard deviation of these means?

14.4 In order to check on the movement of ordinary share prices in *The Times*, a random sample of 30 shares is to be selected from amongst those listed under the heading 'Commercial, Industrial' in the Business News section. Carry out such a random selection and compile a share index for these shares over the period of a month. Plot a graph giving your index, as well as *The Times* index over the same period. (Make the starting index values the same.) Comment on the reasons for any divergencies.

14.5 Use table 13.2 and table 14.1 to draw 50 random samples of size two from a unit normal distribution. For each sample calculate the range (i.e. higher minus lower value). For the 50 values of range so obtained calculate the mean and standard deviation and compare with the theoretical values (1·13 and 0·85 respectively).

15

TIME SERIES

15.1 Many of the functions and quantities studied in the earlier chapters can in practice be measured not just once but repeatedly. A series of values over a period of time is called a *time series*. As a first step in the study of such series some form of diagrammatic representation is very often valuable and enables changes to be detected very swiftly. The usual method is to place time on the horizontal scale and the quantity that is being measured on the vertical scale, but it should be noted that there are in fact two basic types of time series. In the first, a series of measurements relating to some quantity are made at particular instants of time, such as the height of the barometer at Greenwich at noon each day, or the population of England and Wales at 30 June each year. In the second, the measurements are the aggregate amounts in a time interval of some particular commodity, for example the output of cars per month from a factory, or the yield of milk per week from a herd of cows. In graphing such figures it is customary to put the measurement against the exact point of time in the first case, and against the middle point of the time interval concerned in the second.

Table 15.1. *Population of England and Wales (thousands)*

Year	Population	Year	Population
1801	9,000	1881	26,000
1811	10,200	1891	29,000
1821	12,000	1901	32,500
1831	13,900	1911	36,100
1841	15,900	1921	37,900
1851	17,900	1931	40,000
1861	20,100	1941	42,000
1871	22,700	1951	43,800
		1961	46,100

15.2 Tables 15.1 to 15.5 provide some examples of the kind of series that might be met with in practice. Table 15.1 gives the population of England and Wales over 160 years. Table 15.2 gives the total value of money deposited in the clearing banks for a period of 20 years between the two wars. Table 15.3 gives the total annual rainfall in London for the 40 years between 1873 and 1912 inclusive, and is taken from a paper by D. Brunt in the *Philosophical Transactions of the Royal Society*, 1925. Table 15.4 gives the average monthly prices of eggs in England for the years 1934–7

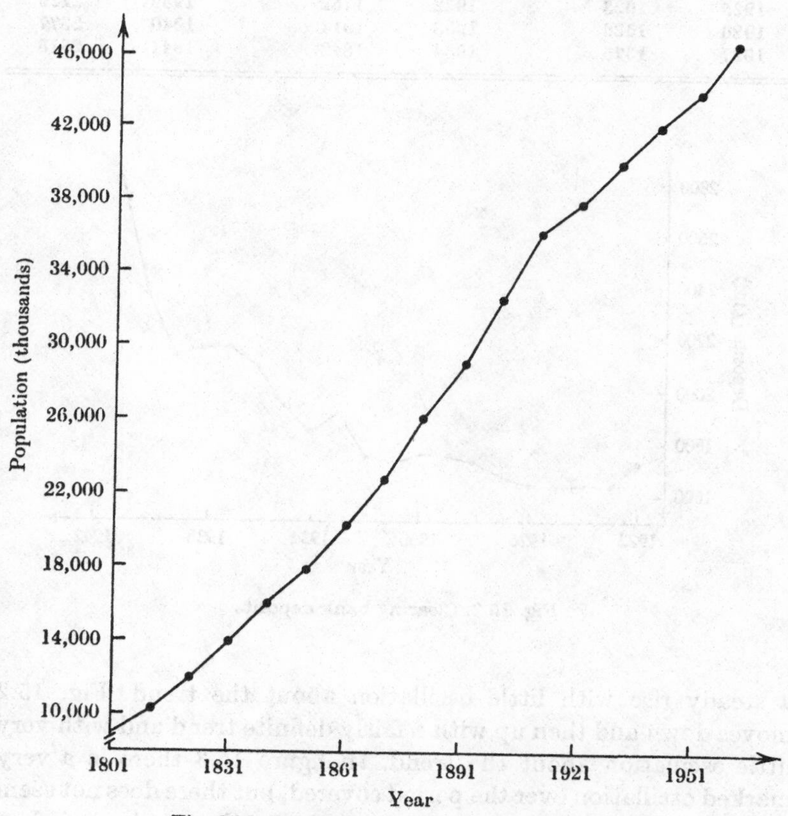

Fig. 15.1. Population of England and Wales

inclusive, and table 15.5 gives the number of persons insured under the National Insurance Scheme in Great Britain who were away from work owing to sickness in each month of the four years 1952–5 inclusive. These tables are illustrated in figs. 15.1 to 15.5 inclusive and various salient features are brought out. Fig. 15.1 illustrates

Table 15.2. *Clearing bank deposits*

Year	Deposits (£m.)	Year	Deposits (£m.)	Year	Deposits (£m.)
1921	1768	1928	1729	1935	1961
1922	1727	1929	1762	1936	2104
1923	1631	1930	1763	1937	2172
1924	1632	1931	1723	1938	2161
1925	1623	1932	1752	1939	2129
1926	1626	1933	1914	1940	2377
1927	1675	1934	1842	1941	2818

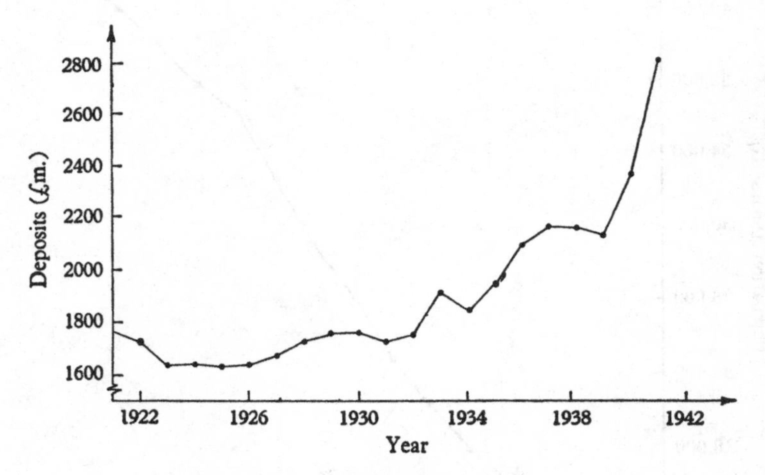

Fig. 15.2. Clearing bank deposits

a steady rise with little oscillation about the trend. Fig. 15.2 moves down and then up with a fairly definite trend and with very little oscillation about the trend. In figure 15.3 there is a very marked oscillation over the period covered, but there does not seem to be very much of a trend apparent over the whole period of 40 years. Figs. 15.4 and 15.5 both indicate very marked seasonal variations that are fairly uniform in their pattern. In fig. 15.4 there is a slight upward trend over the whole period whereas in fig. 15.5 no such trend is apparent. Of course it is possible, though unlikely, that there are other oscillations which are being masked in table 15.2 because the data are only available once a year.

Table 15.3. *Rainfall at London*

Year	Rainfall (in.)	Year	Rainfall (in.)	Year	Rainfall (in.)	Year	Rainfall (in.)	Year	Rainfall (in.)
1873	22·67	1881	27·92	1889	23·85	1897	22·86	1905	22·97
1874	18·82	1882	27·14	1890	21·23	1898	17·69	1906	24·26
1875	28·44	1883	24·40	1891	28·15	1899	22·54	1907	23·01
1876	26·16	1884	20·35	1892	22·61	1900	23·28	1908	23·67
1877	28·17	1885	26·64	1893	19·80	1901	22·17	1909	26·75
1878	34·08	1886	27·01	1894	27·94	1902	20·84	1910	25·36
1879	33·82	1887	19·21	1895	21·47	1903	38·10	1911	24·79
1880	30·28	1888	27·74	1896	23·52	1904	20·65	1912	27·88

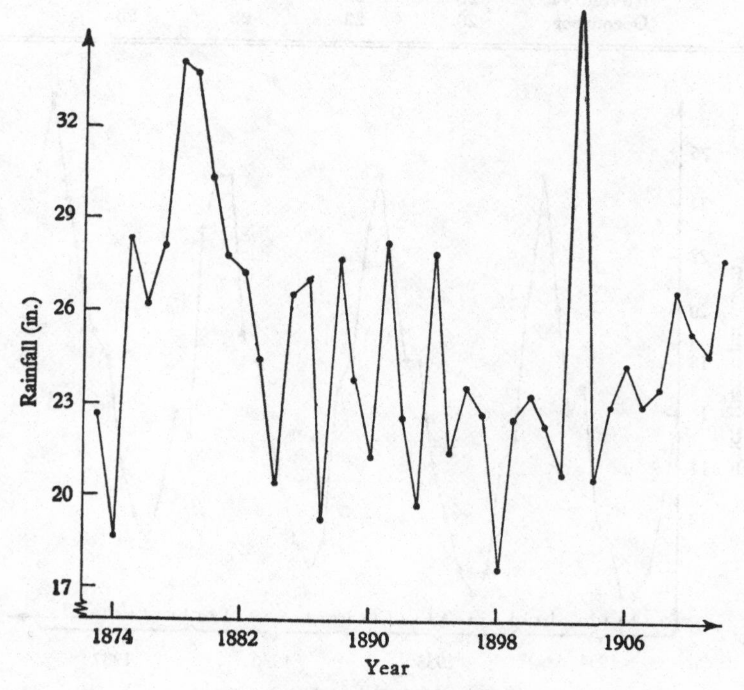

Fig. 15.3. Rainfall at London

15.3 Looking at the tables and figures given here it will be seen that three kinds of movements occur. First there is a general or overall growth which may be up or down, slow or fast, but is broadly a smooth development of the whole system over the years.

Table 15.4. *Average wholesale monthly prices of eggs in England (in pence per dozen)*

Year	1934	1935	1936	1937
January	16	15	18	15
February	13	14	17	16
March	10	10	11	13
April	9	9	10	11
May	9	10	11	12
June	11	12	13	14
July	12	14	15	18
August	17	18	18	19
September	16	18	19	21
October	20	21	25	24
November	25	25	25	28
December	20	23	23	25

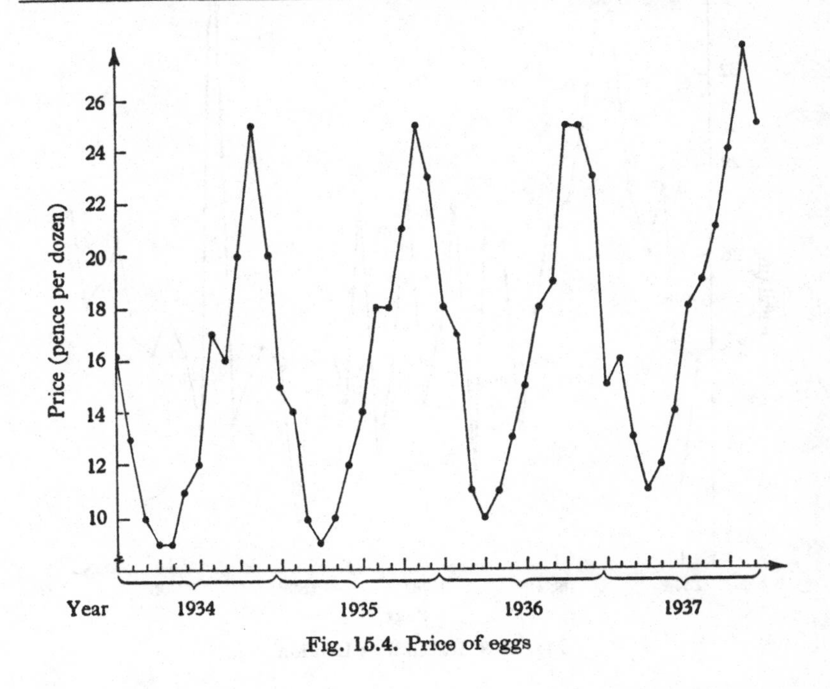

Fig. 15.4. Price of eggs

Thus the population of England and Wales shown in table 15.1 has developed over 160 years from some 9 million in 1801 to 32·5 million in 1901 and 46·1 million in 1961. Although the rate of growth has varied from time to time the general overall picture is of a smooth

Table 15.5. *Insured persons absent from work due to sickness (in units of a thousand)*

Year	1964	1965	1966	1967
January	1,092	1,094	1,177	1,015
February	1,059	1,090	1,170	1,010
March	1,045	1,127	1,168	995
April	992	913	929	944
May	861	974	892	927
June	889	971	849	904
July	851	914	827	879
August	855	916	835	860
September	886	949	835	889
October	980	1,008	876	947
November	1,010	996	934	963
December	929	996	954	997

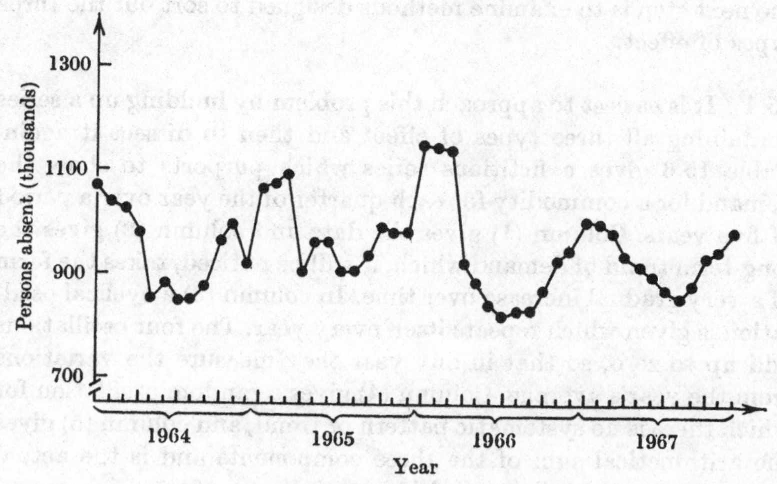

Fig. 15.5. Insured persons absent from work due to sickness

increase all the while. Secondly, superimposed on the regular long-term trend, a time series is often influenced by a group of causes which are not functioning continuously but periodically. For example, due to the usual seasonal changes, the demand for electricity each year is high in the winter and low in the summer. Similarly, in economic data there are often fluctuations over a longer period, referred to as the trade cycle.

Thirdly, there are still further variations that occur when both the long-term trend and the seasonal fluctuations have been removed. These variations may be due to a multiplicity of causes, strikes, floods, wars, accidents and so on. The effects are irregular and of varying magnitude and are therefore called unsystematic, or random, effects. The rainfall data given in table 15.3 appear (fig. 15.3) to consist mainly of random oscillations with little trend or cyclical effect.

Thus any time series can be looked upon as the sum of three different types of effect:

(*a*) general or long-term trend;
(*b*) seasonal or cyclical oscillations;
(*c*) unsystematic or random oscillations.

Any particular series may contain only one or two of these constituents, but there will be cases which contain all of them and the next step is to examine methods designed to sort out the three types of effect.

15.4 It is easiest to approach this problem by building up a series containing all three types of effect and then to dissect it again. Table 15.6 gives a fictitious series which purports to show the demand for a commodity for each quarter of the year over a period of five years. Column (1) gives the date and column (2) gives the long-term trend of demand which, it will be noticed, takes the form of a very gradual increase over time. In column (3) a cyclical oscillation is given which repeats itself every year. The four oscillations add up to zero, so that in any year they measure the variations from the year's average. Column (4) gives a random oscillation for which there is no systematic pattern or trend, and column (5) gives the arithmetical sum of the three components and is the actual series that would be observed in practice.

To examine the resulting series a first step would be to draw a graph of it; this is done in fig. 15.6, where a specimen estimated trend line has been drawn in. This has been done by eye and no doubt different people would all draw a slightly different trend line for the same set of data. From the trend line drawn, there has been a rise of $+3{\cdot}9$ over the five years compared with an actual rise of some $+3{\cdot}8$. The difference is very small but the amount does depend to some extent on personal biases in drawing the line. In earlier chapters it has been demonstrated how the mean of a

Table 15.6. *Constructed series*

(1) Time Year	Quarter	(2) Long-term trend	(3) Seasonal oscillations	(4) Random oscillations	(5) Total series
1	1	100·0	+1·2	+0·2	101·4
	2	100·2	−0·8	+0·6	100·0
	3	100·4	−1·4	−1·0	98·0
	4	100·6	+1·0	−0·6	101·0
2	1	100·8	+1·2	−0·2	101·8
	2	101·0	−0·8	−0·2	100·0
	3	101·2	−1·4	−1·0	98·8
	4	101·4	+1·0	+0·2	102·6
3	1	101·6	+1·2	+0·3	103·1
	2	101·8	−0·8	−0·5	100·5
	3	102·0	−1·4	+0·6	101·2
	4	102·2	+1·0	+0·5	103·7
4	1	102·4	+1·2	−0·2	103·4
	2	102·6	−0·8	−0·7	101·1
	3	102·8	−1·4	+0·4	101·8
	4	103·0	+1·0	+0·6	104·6
5	1	103·2	+1·2	−0·8	103·6
	2	103·4	−0·8	+0·2	102·8
	3	103·6	−1·4	−0·0	102·2
	4	103·8	+1·0	−0·9	103·9

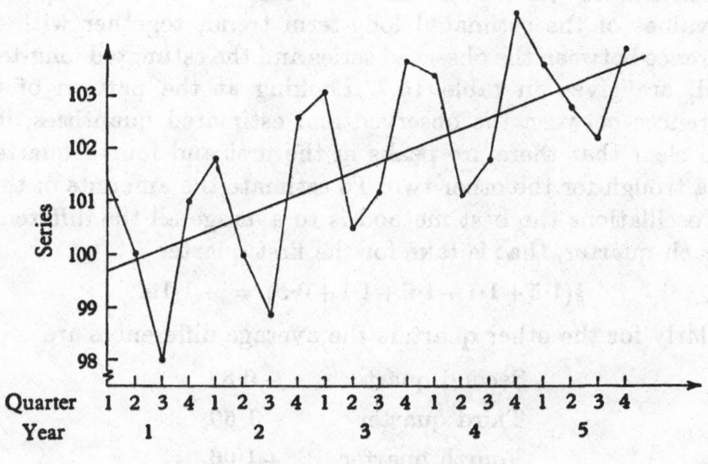

Fig. 15.6. Constructed series

number of quantities increases in accuracy as the number of the quantities used to calculate the mean increases. Further, it would be surmised from the above figures that there is some seasonal variation present, so suppose that, instead of considering the figures for each quarter separately, the average figures for each year were considered. These figures would contain the four quarters of the year once only, and thus any seasonal effect would be eliminated. The values are found to be:

Year 1 100·1,
Year 2 100·8,
Year 3 102·1,
Year 4 102·7,
Year 5 103·1.

Allowing for the fact that these are mid-year estimates only, and thus span four years and not five, the trend is seen to be 3·0 in four years, which by simple proportion is equivalent to about 3·8 in five years, hence verifying what is known to be the underlying trend. To determine the starting-point of the series it is probably easiest to find the middle point first, by averaging all twenty values. This comes to be 101·8 and is a little lower than the true trend value of 101·9 since the random oscillations have tended on average to be slightly more negative than positive. Using the trend, just established, of +3·8 over the five years or +0·2 per quarter, the values of the estimated long-term trend, together with the difference between the observed series and the estimated long-term trend, are given in table 15.7. Looking at the pattern of the differences between the observed and estimated quantities, it is quite clear that there are peaks in the first and fourth quarters, and a trough for the other two. To estimate the amounts of these four oscillations the best method is to average all the differences for each quarter, that is take for the first quarter

$$\tfrac{1}{5}(1{\cdot}5+1{\cdot}1+1{\cdot}6+1{\cdot}1+0{\cdot}5) = +1{\cdot}16.$$

Similarly for the other quarters the average differences are

Second quarter −0·82,
Third quarter −1·50,
Fourth quarter +1·06.

A comparison of the true oscillations (that is the oscillations origin-

Table 15.7. *Estimated trend*

Time		Estimated	Difference	Time		Estimated	Difference
Year	Quarter	trend	Obs. − Est.	Year	Quarter	trend	Obs. − Est.
1	1	99·9	+1·5	3	3	101·9	−0·7
	2	101·1	−0·1		4	102·1	+1·6
	3	100·3	−2·3	4	1	102·3	+1·1
	4	100·5	+0·5		2	102·5	−1·4
2	1	100·7	+1·1		3	102·7	−0·9
	2	100·9	−0·9		4	102·9	+1·7
	3	101·1	−2·3	5	1	103·1	+0·5
	4	101·3	+1·3		2	103·3	−0·5
3	1	101·5	+1·6		3	103·5	−1·3
	2	101·7	−1·2		4	103·7	+0·2

ally used to build up the series) with the calculated oscillations, gives the figures in table 15.8 and shows a very close agreement. Notice that the four calculated oscillations add up to −0·10 and hence in practice one would adjust all these figures by a quantity +0·025 in order to make it a pure oscillation.

Table 15.8. *Calculated seasonal oscillations*

	True	Calculated	Error
First quarter	+1·20	+1·16	+0·04
Second quarter	−0·80	−0·82	+0·02
Third quarter	−1·40	−1·50	+0·10
Fourth quarter	+1·00	+1·06	−0·06

15.5 In the example just discussed the period of the seasonal observations with their cycle of four observations was very marked, and could be picked out without difficulty once the overall trend had been eliminated. This will not always be the case, however, and methods must be devised to enable the period of the oscillations, as well as their size, to be established. But it is necessary first to find more general methods for the determination of the trend over the whole period. The principle adopted is based on the fact that if new terms containing one of each of the cyclical observations are formed, the cyclical oscillations, which appear equally in each term, are eliminated. Instead, therefore, of considering the observations by themselves, a series of averages is formed, where each average contains a number of successive observations. These averages are called successive or *moving averages*. The number of points in a moving average is the number of items that are averaged. Thus a 2-point moving average gives the average of two successive observations and this is repeated for the whole series of observations. The process is illustrated, for the data of the previous example, in table 15.9. Column (3) gives a 2-point moving average. Thus $\frac{1}{2}(101{\cdot}4+100{\cdot}0) = 100{\cdot}7$, which is placed half-way between the two values concerned. Then the next value will be

$$\tfrac{1}{2}(100{\cdot}0+98{\cdot}0) = 99{\cdot}0,$$

placed again half-way between the two values concerned. The procedure is repeated all the way through the series and produces twenty-three values instead of the original twenty-four as the series now starts and stops half a unit from each end value. If these 2-point moving averages are plotted it will be found that there is still no really smooth trend apparent and quite a number of oscillations. Although the series now starts at about 100 and ends at 103, there are nine occasions when a value in the series is less than the value preceding it and this is still contrary to the trend, which is always in an increasing direction. Column (4) gives a 3-point moving average. The calculations are similar to those of the 2-point average. Thus the first three values are

$$\tfrac{1}{3}(101{\cdot}4+100{\cdot}0+98{\cdot}0) = 99{\cdot}8,$$
$$\tfrac{1}{3}(100{\cdot}0+98{\cdot}0+101{\cdot}0) = 99{\cdot}7,$$
$$\tfrac{1}{3}(98{\cdot}0+101{\cdot}0+101{\cdot}8) = 100{\cdot}3,$$

and each value is placed opposite the middle value of the three that have been used. This series produces fewer oscillations than did the 2-point moving average, but the oscillations have not by any means been completely eliminated and the series does not progress uniformly upwards. The 4-point moving average in column (5) produces a series which, with three small exceptions, increases as the time progresses. This is to be expected since, by taking averages of four successive observations, each quarter appears once, and if there is any quarterly effect it will appear equally in each item of the resulting series of observations. When a 5-point moving average is calculated there is once again a lack of uniform trend in the resulting observations. This demonstrates that there is an optimum number of points in a moving average

Table 15.9. *Calculation of moving averages*

(1) Time		(2)	(3)	(4)	(5)	(6)
Year	Quarter	Observation	2-point moving average	3-point moving average	4-point moving average	5-point moving average
1	1	101·4				
			100·7			
	2	100·0		99·8		
			99·0		100·1	
	3	98·0		99·7		100·4
			99·5		100·2	
	4	101·0		100·3		100·2
			101·4		100·2	
2	1	101·8		100·9		99·9
			100·9		100·4	
	2	100·0		100·2		100·8
			99·4		100·8	
	3	98·8		100·5		101·3
			100·7		101·1	
	4	102·6		101·5		101·0
			102·8		101·2	
3	1	103·1		102·1		101·2
			101·8		101·8	
	2	100·5		101·6		102·2
			100·8		102·1	
	3	101·2		101·8		102·4
			102·4		102·2	
	4	103·7		102·8		102·0
			103·5		102·3	
4	1	103·4		102·7		102·2
			102·2		102·5	
	2	101·1		102·1		102·9
			101·4		102·7	
	3	101·8		102·5		102·9
			103·2		102·8	
	4	104·6		103·3		102·8
			104·1		103·2	
5	1	103·6		103·7		103·0
			103·2		103·3	
	2	102·8		102·9		103·4
			102·5		103·1	
	3	102·2		103·0		
			103·0			
	4	103·9				

and an increase in the number of points will not necessarily result in a smoother series. Of course, in this example any number of points that is a multitude of four, for instance eight, would also produce a smooth series.

15.6 The purpose of using the moving average has been twofold. First, provided a suitable number of points are used, it enables the seasonal or cyclical oscillations to be ironed out, and the trend is thus left more exposed. Secondly, by taking averages the effects of any random fluctuations are automatically reduced and hence do not swamp the main effects. However, it should be noticed that even if these effects did not in fact exist in the original series, nothing will be lost by carrying out the averaging procedure, which will produce a series very similar to the original one. This is, of course, the desired function of the method, but it is important to note that only if it is a linear trend does the method reproduce the trend exactly. To illustrate this consider a series produced by giving x the successive values 1, 2, 3, ... in the expression $y = (x-5)^2$. The first nine values of this series, together with the calculation of the moving averages for three different numbers of points are given in table 15.10. From this table the original series and 3- and 5-point moving averages have been plotted in fig. 15.7. It will be noticed that although the curves for the averages are very similar to the curve of the original series of observations, the values are all above them and the greater the number of points the greater the departure of the average from the original series. This unfortunate property of moving averages, namely that with a concave series of observations a moving average tends to overestimate the trend effect, must be borne in mind. It will be found that with a series of observations that is convex the situation is reversed, and the moving average produces values of trend that are too low. In a series that is partly of one form and partly of the other the values produced would sometimes lean one way and sometimes the other. This particular case is illustrated in fig. 15.8 using the data of table 15.11 in which the values of the simple algebraic function

$$y = (\tfrac{1}{2}x-3)^3+20$$

are given for integral values of x from 1 to 11. In the table the moving averages using 2, 3 and 4 points can be seen to lag behind the true trend values at the beginning when the curve is convex, but to be ahead of the trend for the latter part of the series when the curve is concave. The curves in fig. 15.8 illustrate this difference and the distortion of curved trends must be remembered when analysing such series.

Table 15.10. *Moving averages $y = (x-5)^2$*

x	y	3-point moving averages	4-point moving averages	5-point moving averages
1	16			
2	9	9·7		
			7·5	
3	4	4·7		6·0
			3·5	
4	1	1·7		3·0
			1·5	
5	0	0·7		2·0
			1·5	
6	1	1·7		3·0
			3·5	
7	4	4·7		6·0
			7·5	
8	9	9·7		
9	16			

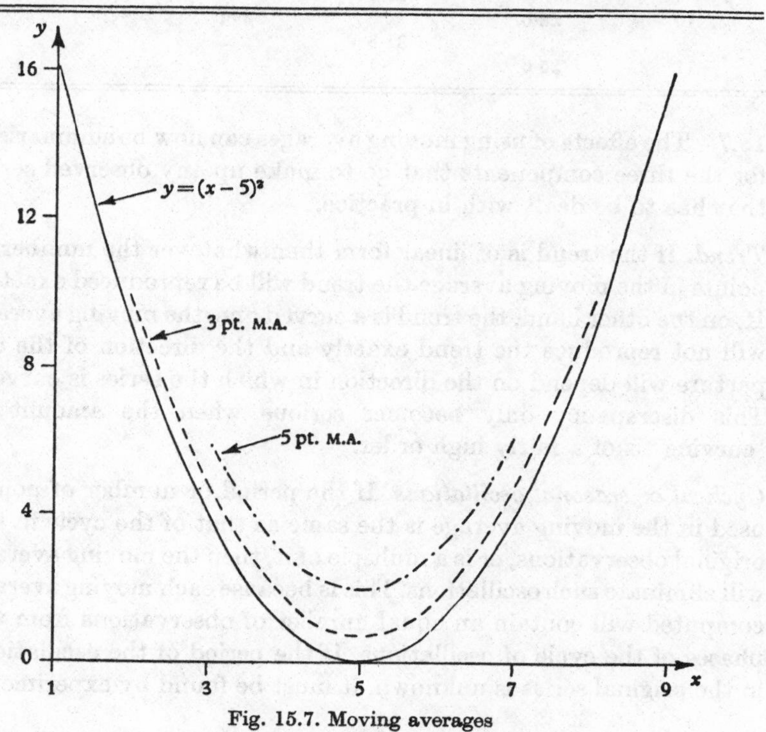

Fig. 15.7. Moving averages

Table 15.11. *Effect of moving averages*

x	y	2-point moving averages	3-point moving averages	4-point moving averages
1	4·4			
		8·2		
2	12·0		11·0	
		14·3		13·00
3	16·6		15·9	
		17·8		16·88
4	19·0		18·5	
		19·45		18·87
5	19·9		19·6	
		19·95		19·75
6	20·0		20·0	
		20·05		20·25
7	20·1		20·4	
		20·55		21·12
8	21·0		21·5	
		22·2		23·12
9	23·4		24·1	
		25·7		27·00
10	28·0		29·0	
		31·8		
11	35·6			

15.7 The effects of using moving averages can now be summarised for the three components that go to make up any observed series that has to be dealt with in practice.

Trend. If the trend is of linear form then whatever the number of points in the moving average the trend will be reproduced exactly. If, on the other hand, the trend is a curved one, the moving average will not reproduce the trend exactly and the direction of the departure will depend on the direction in which the series is curved. This discrepancy only becomes serious when the amount of 'curving' is of a fairly high order.

Cyclical or seasonal oscillations. If the period or number of points used in the moving average is the same as that of the cycle in the original observations, or is a multiple of it, then the moving average will eliminate such oscillations. This is because each moving average computed will contain an equal number of observations from the phases of the cycle of oscillations. If the period of the oscillations in the original series is unknown, it must be found by experiment-

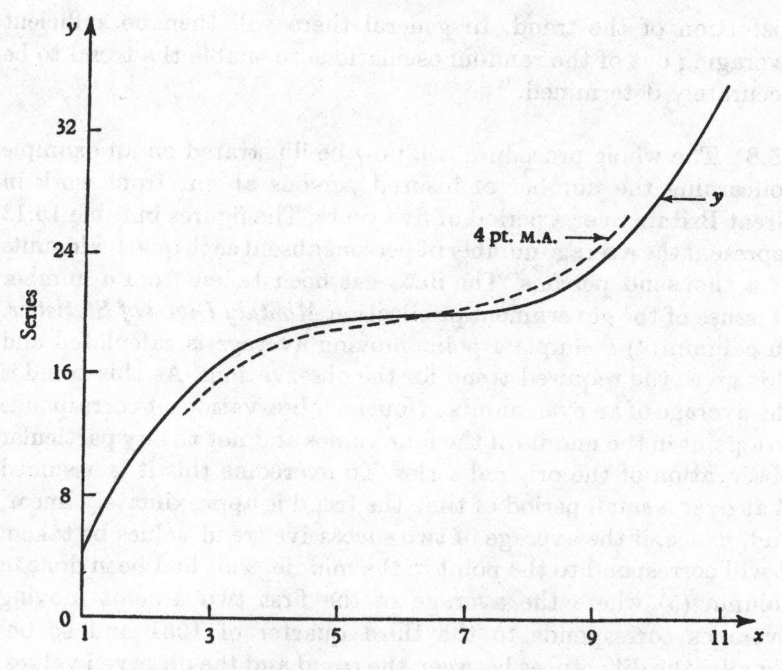

Fig. 15.8. Curvature effect with $y = (\frac{1}{2}x - 3)^3 + 20$

ing with different numbers of points for the moving average and finding the number of points which produces the smoothest form of series for the trend.

Random oscillations. Since these operate sometimes positively and sometimes negatively, the more individuals included in the moving average the more are the oscillations likely to be eliminated, or at any rate reduced to such a magnitude that they have little practical effect on the determination of the trend. Thus, if the cyclical oscillations have only a very small period, the random oscillations may affect the moving average with the correct number of points, but when multiples of the correct number of points are reached the random oscillations will have been much further damped down and should not affect the determination of the trend. In practice the number of points taken for the moving average will be the number corresponding to the cyclical oscillations and it is to be hoped that this moving average will produce no significant distortion of the trend. In general there will then be sufficient averaging out of the random oscillations to enable the trend to be accurately determined.

15.8 The whole procedure will now be illustrated on an example concerning the number of insured persons absent from work in Great Britain over a period of five years. The figures in table 15.12 represent the average number of persons absent each quarter in units of a thousand persons. The data has been taken from a number of issues of the government publication *Monthly Digest of Statistics.* In column (4) a simple 4-point moving average is calculated and this gives the required trend for the observations. As this trend is the average of an even number (four) of observations it corresponds to a point in the middle of the four values and not to any particular observation of the original series. To overcome this it is assumed that over a small period of time the trend is approximately linear, and, hence, if the average of two successive trend values be taken, it will correspond to the point in the middle. This had been done in column (5) where the average of the first two 4-point moving averages corresponds to the third quarter of 1951 and so on. Finally, the differences between the trend and the observed values, shown in column (6), are an indication of the seasonal fluctuations. The values of these deviations are rearranged in table 15.13 to correspond with the four quarters of the year and the very distinct pattern over the years is made apparent. These quarterly deviations still include a fair amount of random oscillation and obviously the more quarterly deviations included, the greater will be the accuracy of any estimate of the quarterly oscillations. Thus in this case the best estimate of the oscillation due to each quarter will be found by taking the average of the four values obtained for each oscillation. These values are given at the bottom of table 15.13 and it will be noticed that they approximately sum to zero. The slight discrepancy is due to the fact that the series is of finite length and the end values are not included in as many of the moving averages as are the central values. It is possible to make a slight adjustment to force the quarterly oscillations to sum to zero but, unless the series is a long one, it is not usually worth making such an adjustment. The original series may now be broken up into the three components, namely trend, seasonal oscillations and random oscillations. This is done in table 15.14 where the random oscillations in column (5)

Table 15.12. *Insured persons absent from work*

(1) Year	(2) Quarter	(3) Sickness nos. (000's)	(4) 4-point moving averages	(5) Trend	(6) Deviation
1951	1	1,170			
	2	833			
			906·50		
	3	781		879·0	+98·0
			851·50		
	4	842		845·7	+3·7
			840·00		
1952	1	950		839·1	−110·9
			838·25		
	2	787		844·9	+57·9
			851·50		
	3	774		871·5	+97·5
			891·50		
	4	895		904·6	+9·6
			917·75		
1953	1	1,110		924·4	−185·6
			931·00		
	2	892		932·7	+40·7
			943·50		
	3	827		928·4	+101·4
			922·25		
	4	909		922·9	+13·9
			923·50		
1954	1	1,061		922·7	−138·3
			922·00		
	2	897		922·1	+25·1
			922·25		
	3	821		925·5	+104·5
			928·75		
	4	910		927·9	+17·9
			927·00		
1955	1	1,087		926·4	−160·6
			925·75		
	2	890		923·1	+33·1
			920·50		
	3	816			
	4	889			

are found in such a manner that the original observations in column (6) are reproduced, that is the identity

trend + seasonal oscillation + random oscillation = observation,

or

$$(3)+(4)+(5) = (6),$$

Table 15.13. *Quarterly deviations*

Quarter	1	2	3	4
1951	—	—	+98·0	+3·7
1952	−110·9	+57·9	+97·5	+9·6
1953	−185·6	+40·7	+101·4	+13·9
1954	−138·3	+25·1	+104·5	+17·9
1955	−160·6	+33·1	—	—
Total	−595·4	+156·8	+401·4	+45·1
Average	−148·9	+39·2	+100·4	+11·3

Table 15.14. *Composition of observations*

(1) Year	(2) Quarter	(3) Trend	(4) Seasonal oscillation	(5) Random oscillation	(6) Observation
1951	3	879·0	−100·4	+2·4	781
	4	845·7	−11·3	+7·6	842
1952	1	839·1	+148·9	−38·0	950
	2	844·9	−39·2	−18·7	787
	3	871·5	−100·4	+2·9	774
	4	904·6	−11·3	+1·7	895
1953	1	924·4	+148·9	+36·7	1,110
	2	932·7	−39·2	−1·5	892
	3	928·4	−100·4	−1·0	827
	4	922·9	−11·3	−2·6	909
1954	1	922·7	+148·9	−10·6	1,061
	2	922·1	−39·2	+14·1	897
	3	925·5	−100·4	−4·1	821
	4	927·9	−11·3	−6·6	910
1955	1	926·4	+148·9	+11·7	1,087
	2	923·1	−39·2	+6·1	890

is satisfied for each observation. Looking down the column of random oscillations the figures are sometimes positive and sometimes negative, in fact eight have each sign. They range in magnitude from 1·0 to 38·0 and there does not seem to be any tendency for particular values to be associated with particular seasons. This is a desirable feature, as otherwise it would show that the seasonal oscillations have been wrongly deduced. The position is illustrated in fig. 15.9 where the original series and the trend of column (3) are plotted. From this it will be seen that the trend is a concave curve with an upward tendency at the start and then is almost horizontal for the end part of the series. This seems to indicate that it should be perfectly possible to make a reasonable forecast of absences for sickness for a few months ahead, but it should be borne in mind how very full of pitfalls any form of forecasting can be in practice. Thus a bad spell of weather can easily cause a 'flu' epidemic to spread very wildly and raise the sickness

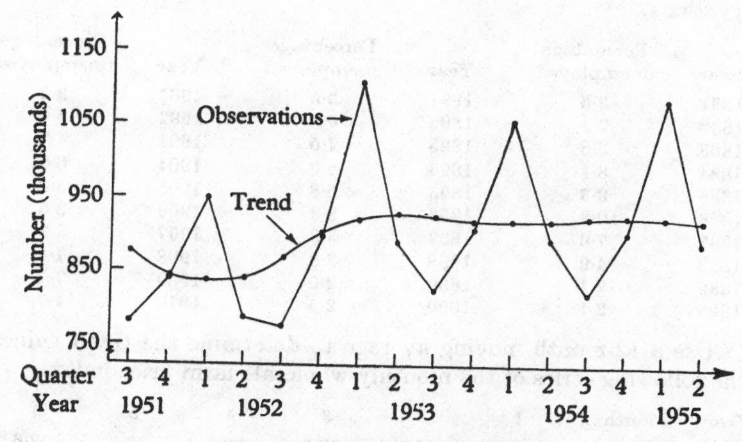

Fig. 15.9. Insured persons absent

rate enormously, giving results well above those that seem to be likely from the study of the series of results available. The main use of an analytical study of any time series, such as is attempted here, is for short-term predictions, and for forecasts of likely seasonal or cyclical variations. This can be useful in planning so that the excess or loss produces no hardship, and it also enables those in positions of responsibility to be able to determine when a significant departure from established levels has taken place. In many forms of economic data, where the theory of time series is often applied, there are a large number of variables in operation which do not act fully independently of each other. This may be unknown to the worker faced with a series of observations of one variable, and some unknown administrative decision about another variable not being examined can affect the series of variables under observation. This interdependence of two variables will be taken up in the next chapter and it is sufficient to say here that in any time series a sudden jump in the observed series should always be investigated to see whether any external causes can be found.

EXERCISES

15.1 The following data give the yearly average percentage of unemployed in certain industries over thirty years. By using moving averages find a suitable period and hence deduce the trend of the observations.

Year	Percentage unemployed	Year	Percentage unemployed	Year	Percentage unemployed
1881	3·5	1891	3·5	1901	3·3
1882	2·3	1892	6·3	1902	4·0
1883	2·6	1893	7·5	1903	4·7
1884	8·1	1894	6·9	1904	6·0
1885	9·3	1895	5·8	1905	5·0
1886	10·2	1896	3·3	1906	3·6
1887	7·6	1897	3·3	1907	3·7
1888	4·9	1898	2·8	1908	7·8
1889	2·1	1899	2·0	1909	7·7
1890	2·1	1900	2·5	1910	4·7

15.2 Use a 12-month moving average to determine the trend values for the following series of the monthly wholesale farm price index.

Year	Month ...	1	2	3	4	5	6
1942	Index	96·0	96·7	97·6	98·7	98·8	98·6
1943	Index	101·9	102·5	103·4	103·7	104·1	103·8
1944	Index	103·3	103·6	103·8	103·9	104·0	104·3
1945	Index	104·9	105·2	105·3	105·7	106·0	106·1

Year	Month ...	7	8	9	10	11	12
1942	Index	98·7	99·2	99·6	100·0	100·3	101·0
1943	Index	103·2	103·1	103·1	103·0	102·9	103·2
1944	Index	104·1	103·9	104·0	104·1	104·4	104·7
1945	Index	105·9	105·7	105·2	105·9	106·8	107·1

15.3 Use a 12-point moving average to determine

(*a*) the trend values,

(*b*) the seasonal oscillation,

(*c*) the random oscillation,

for the following figures of mean daily air temperature (in degrees Fahrenheit) at sea-level in England. Find also the mean and standard deviations of the random oscillations found in (*c*) above.

Year	Month ...	1	2	3	4	5	6
1948	Temperature	42·5	41·4	47·8	49·1	53·4	57·3
1949	Temperature	42·6	43·2	42·2	50·9	53·0	59·5
1950	Temperature	40·8	42·9	46·6	46·7	53·3	61·7
1951	Temperature	40·0	39·7	40·8	45·3	51·0	57·5

Year	Month ...	7	8	9	10	11	12
1948	Temperature	60·6	60·0	57·9	51·4	46·4	43·5
1949	Temperature	63·7	63·2	62·4	54·4	45·0	43·7
1950	Temperature	61·5	61·2	56·5	50·9	43·7	35·9
1951	Temperature	62·0	59·9	58·6	50·8	48·0	43·4

15.4 By using a 4-point moving average split up the following series of marriages in the United Kingdom into trend, seasonal and random oscillations. Estimate the number of marriages that will take place in the second quarter of 1956, and compare it with the number that actually took place.

	Quarter			
Year	First	Second	Third	Fourth
1952	121·3	79·8	118·3	80·4
1953	107·6	87·5	120·8	78·9
1954	110·4	85·0	120·2	77·3
1955	115·3	85·6	122·5	86·8

(Figures given are in units of a thousand.)

15.5 The following table gives the index of retail prices of food that are included with other items, in the cost-of-living index. Use the figures to examine whether there is any seasonal fluctuation in the prices of food and, if so, how much.

	Month			
Year	January	April	July	October
1952	100·0	103·9	108·5	108·3
1953	109·2	112·5	113·7	110·6
1954	110·2	112·6	118·0	116·1
1955	119·2	119·9	125·6	125·9
1956	125·4	132·5	126·8	127·4

15.6 The following figures are available for the electricity generated for public supply, but the two spaces marked with an asterisk are not available. Make an estimate of the missing figures, and also one for July 1956.

	Month			
Year	January	April	July	October
1952	6319	4775	4108	5526
1953	6609	5269	4489	*
1954	*	5634	4972	6241
1955	7988	6102	5115	7022
1956	8588	6952	?	—

(Units are million kilowatt hours.)

15.7 The table gives the exports of a certain country over a period of twenty-two years. *X* in writing to a newspaper points out that the exports are declining and quotes the figures:

Average exports	1929–31	£198 m.
	1935–37	£197 m.
	1941–43	£196 m.

Y then replies that the converse is true and quotes the figures:

Average exports	1926–28	£200 m.
	1932–34	£204 m.
	1938–40	£206 m.

Which of them, X, or Y, is really correct and why?

Year	Exports (£m.)	Year	Exports (£m.)	Year	Exports (£m.)
1924	200	1932	205	1939	207
1925	198	1933	205	1940	208
1926	197	1934	202	1941	200
1927	201	1935	199	1942	195
1928	202	1936	195	1943	193
1929	200	1937	197	1944	200
1930	195	1938	203	1945	208
1931	199				

15.8 The following figures give the number of car licences current in Great Britain (in thousands) on certain dates.

Year	28 February	31 May	31 August	30 November
1952	2195	2397	2467	2448
1953	2371	2625	2724	2720
1954	2627	2934	3064	3059
1955	2956	3309	3479	3472
1956	3325	3708	3835	3801

Estimate

(*a*) the trend, using the method of moving averages;

(*b*) the average fluctuations from the trend for each of the four quarters;

(*c*) the mean and standard deviation of the residuals when both trend and quarterly variations have been eliminated.

15.9 The table below gives the average colliery cost per ton of deep-mined coal for each quarter over a period of five years. The figures have been extracted from the National Coal Board quarterly statements.

	Quarter			
	First	Second	Third	Fourth
1953	57·28	59·22	61·16	59·41
1954	59·76	61·49	63·43	63·24
1955	63·22	67·11	70·09	69·15
1956	70·35	73·46	78·17	76·37
1957	76·02	79·30	86·76	85·23

(Unit is shillings per ton of saleable output.)

Estimate the average percentage rate of increase in the cost per ton per year

(*a*) by using a 4-point moving average on the given figures in order to establish the trend;

(*b*) by applying (*a*) to the logarithms of the colliery costs per ton. Which method is the more preferable, and why?

15.10 The following tabulation gives the index of industrial production (1958 = 100) over five years.

Quarter	First	Second	Third	Fourth
1960	115	113	106	116
1961	116	116	107	116
1962	116	117	109	118
1963	115	119	114	127
1964	128	130	121	134

(*a*) Plot the series.

(*b*) Calculate a moving average trend and plot on to (*a*).

(*c*) Estimate the average quarterly fluctuations about the trend for each quarter of the year.

16

PAIRS OF CHARACTERS

16.1 So far the methods of statistical analysis examined have dealt with samples of individuals on each of which a single measurement has been made. The tests concerned have been developed utilising the probability distribution of this single measurement, x. Many statistical problems, however, are concerned with more than a single characteristic of each individual. For instance, the height, x, and weight, y, of a number of schoolboys are recorded and it is, desired to examine the relationship between the two measurements.

Table 16.1. *Length of copper rod*

Temperature in °C. (x)	Length in mm. (y)	Temperature in °C. (x)	Length in mm. (y)
20·4	2461·12	42·9	2462·03
27·3	2461·41	58·3	2462·69
38·5	2461·86	67·4	2463·05

Sometimes the relationship between the two measurements is very marked so that any statistical analysis will be quite straightforward. For example, an examination of the length of a copper bar at various temperatures was carried out in a laboratory under very accurate conditions. The results are given in table 16.1. The most straightforward way of analysing this data is to plot the six pairs of measurements on graph paper, the x-axis denoting the temperature in degrees centigrade and the y-axis the length of the rod in millimetres. This is illustrated in fig. 16.1 and it will be seen that the points lie almost exactly on a straight line: in fact it is quite a simple matter to draw by eye a straight line that very nearly passes through all the six points on the figure. From the figure it is then a reasonably straightforward matter to make a quick and accurate estimate of the length of the rod for any intermediate temperature. For instance, if the temperature were 61·2° C. the dotted line in the figure indicates that the corresponding point on the line gives a length of 2462·81 mm. Alternatively, if the rod measured 2461·93 mm. the temperature would be approximately 40·3° C. Thus the line can be used for determinations of temperature for a given length and of length for a given temperature. Such a line is called a *regression line*. If a line is used to estimate y from x it is referred to as the regression line of y on x.

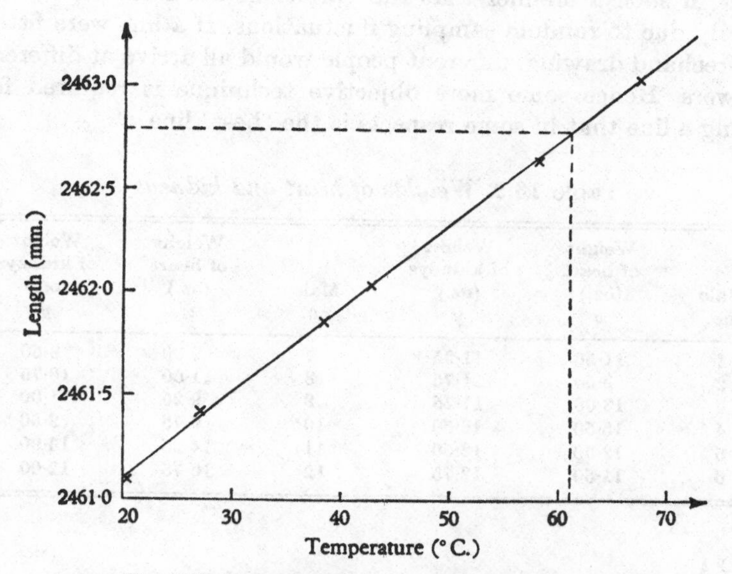

Fig. 16.1. Temperature and length of copper rod.

If the line is used to estimate x from y then it is referred to as the regression line of x on y. In the example just discussed the points lay so closely on a straight line that there could be only one regression line whether y was being estimated from x or the other way round. Later examples will show, however, that this is not always the case and there may be two lines to be considered.

16.2 In many practical problems the pairs of measurements obtained do not lie quite so obviously along a straight line as in the case of the copper bar.

Example 16.1 Table 16.2 gives the weight of heart, x, and the weight of kidneys, y, in a selected set of twelve adult males between the ages of 25 and 55 years. These are plotted in fig. 16.2 and it will be seen that a unique straight line can no longer be drawn through the twelve observations. If there is believed to be an underlying linear relationship between the two variables, the fact that there is no such unique straight line must be due to random fluctuations in the individuals selected for measurement. The 'best' line would, therefore, go through the middle of the observations in such a manner that the variations about the line were merely due to random sampling fluctuations. If a line were fitted by freehand drawing, different people would all arrive at different answers. Hence some more objective technique is required for fitting a line that in some respects is the 'best' line.

Table 16.2. *Weights of heart and kidneys*

Male no.	Weight of heart (oz.) x	Weight of kidneys (oz.) y	Male no.	Weight of heart (oz.) x	Weight of kidneys (oz.) y
1	11·50	11·25	7	9·00	9·50
2	9·50	11·75	8	11·50	10·75
3	13·00	11·75	9	9·25	11·00
4	15·50	12·50	10	9·75	9·50
5	12·50	12·50	11	14·25	13·00
6	11·50	12·75	12	10·75	12·00

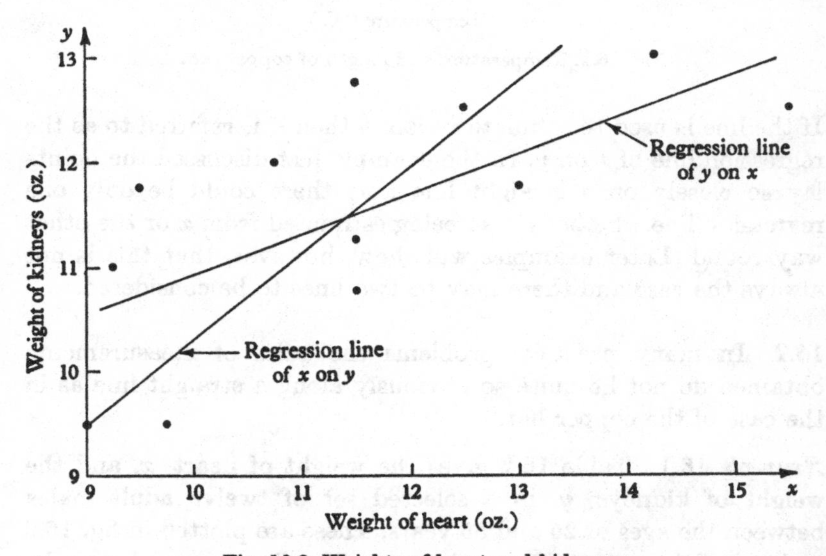

Fig. 16.2. Weights of heart and kidneys

The first step is to decide which variable is to be estimated from the other. In this case it will be assumed that kidney weight is to be estimated from the heart weight. For any fitted line the difference between the observed value of y and the corresponding value of y on the line can be found. These differences are squared and added up for all the observations and the fitted line is chosen so that the sum of these squared differences is a minimum. Notice that the differences themselves cannot be minimised since some would be positive and some negative and a line could appear to be a good fit from the low sum of differences yet was actually a bad fit to the observations. If these squared differences are looked upon as the errors in estimating y from a set of values of x, then the procedure consists in minimising the sum of squares of the errors. It is, however, not necessary to draw a number of possible lines by eye, measure the errors for each of the points, and finally find the line that gives the minimum sum of squared errors. There is a straightforward algebraic method for finding the required line.

Suppose that, in general, there are n pairs of measurements (x, y). It is known from the plotting of graphs that any straight line relationship between x and y is of the form $y = bx+c$, where b and c are constants. Thus if x is equal to 2 the corresponding value of y is $2b+c$ and if x is equal to 3·5 the corresponding value of y is $3{\cdot}5b+c$. b is called the slope of the line and measures its steepness. If b is zero the line is horizontal, whilst if b is very large the line is almost vertical. For a regression line the form of the equation is

$$y = bx + (\bar{y} - b\bar{x}),$$

or

$$y - \bar{y} = b(x - \bar{x}), \qquad (16.1)$$

where $\bar{y}$ is the mean of the values of y, that is

$$\bar{y} = \Sigma y/n,$$

$\bar{x}$ is similarly the mean of the values of x, that is

$$\bar{x} = \Sigma x/n,$$

whilst b is the slope of the line and depends on the relationship between x and y. The formula is

$$b = \frac{\Sigma(xy) - n\bar{x}\,.\,\bar{y}}{\Sigma(x^2) - n\bar{x}^2}. \qquad (16.2)$$

The symbol $\Sigma(xy)$ stands for the sum of the cross products of x and y; that is, corresponding values of x and y are multiplied together and these resulting products are added up for the n observations. The denominator will be recognised from formula (7.3) as being n times the variance of the values of x. Using the formulae above the computations are relatively straightforward and are shown in table 16.3.

Table 16.3. *Computations for regression line*

Observation	x	x^2	y	y^2	xy
1	11·50	132·25	11·25	126·56	129·37
2	9·50	90 25	11·75	138·06	111·62
3	13·00	169·00	11·75	138·06	152·75
4	15·50	240·25	12·50	156·25	193·75
5	12·50	156·25	12·50	156·25	156·25
6	11·50	132·25	12·75	162·56	146·62
7	9·00	81·00	9·50	90·25	85·50
8	11·50	132·25	10·75	115·56	123·62
9	9·25	85·56	11·00	121·00	101·75
10	9·75	95·06	9·50	90·25	92·62
11	14·25	203·06	13·00	169·00	185·25
12	10·75	115·56	12·00	144·00	129·00
Total	138·00	1632·75	138·25	1607·81	1608·12

From the table the required quantities are

$$\Sigma x = 138{\cdot}00, \qquad \Sigma y = 138{\cdot}25,$$

$$\Sigma xy = 1608{\cdot}12, \qquad \Sigma x^2 = 1632{\cdot}75.$$

The column of y^2 is not needed at this stage but will be used later. The three expressions required for the regression equation are

$$\bar{x} = 11{\cdot}50, \qquad \bar{y} = 11{\cdot}52,$$

$$b = \frac{1608{\cdot}12 - 1589{\cdot}875}{1632{\cdot}75 - 1587} = \frac{18{\cdot}245}{45{\cdot}75} = 0{\cdot}3988.$$

Hence the required equation will be

$$y - 11{\cdot}52 = 0{\cdot}3988(x - 11{\cdot}50).$$

or

$$y = 0{\cdot}3988x + 6{\cdot}9338.$$

This line is plotted in fig. 16.2 and would be used to predict or estimate the kidney weight, y, for any given value of x. Thus for a heart weight of 12·2 the estimated kidney weight would be 11·80. It should be noted that the regression line will always go through the point that corresponds to the mean of the observations of x and y. This can be seen from (16.1) because both the left- and right-hand sides of the equation become zero at the point $(\bar{x}, \bar{y})$. This provides a useful check when drawing any regresssion line.

16.3 Suppose now that the problem was posed the other way round, and that it was desired to estimate x from y. This requires the regression line of x on y, and the form that it takes must clearly be obtainable from (16.1) and (16.2) by interchanging x and y wherever they occur. Thus the regression line of x on y will be

$$x - \bar{x} = b'(y - \bar{y}), \tag{16.3}$$

where

$$b' = \frac{\Sigma xy - n\bar{x}.\bar{y}}{\Sigma(y^2) - n\bar{y}^2}. \tag{16.4}$$

These equations are very similar to the earlier equations and the only new quantity required is $\Sigma(y^2)$, which was obtained in the last but one column of table 16.3. Hence

$$b' = \frac{18{\cdot}245}{15{\cdot}055} = 1{\cdot}2119,$$

and the required regression equation is

$$x - 11{\cdot}50 = 1{\cdot}2119(y - 11{\cdot}52),$$

or

$$x = 1{\cdot}2119y - 2{\cdot}4611.$$

It should be noted that this is not the same equation as before and it will be seen from fig. 16.2 that there is some considerable difference. The reason for this difference is that two different quantities are being minimised, namely

Regression of y on x: sum for each x of (estimated y − observed y)^2.

Regression of x on y: sum for each y of (estimated x − observed x)^2.

What this implies geometrically is demonstrated in fig. 16.3 where (*a*) shows the vertical deviations which are to be minimised in order to find the regression line of y on x whilst (*b*) shows the horizontal deviations which have to be minimised in order to obtain the regression line of x on y. In general, these two processes will not lead to the same answer, unless there is a perfect linear relationship between the two variables.

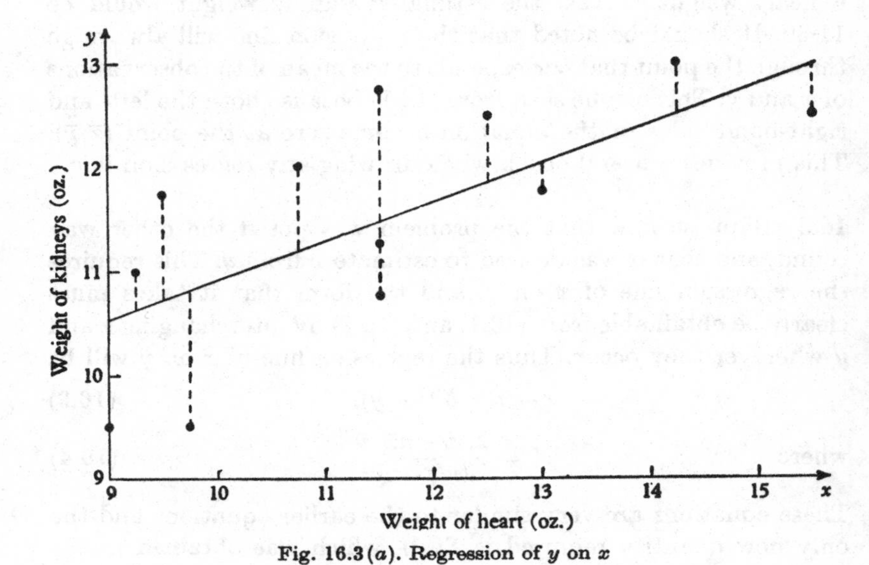

Fig. 16.3 (*a*). Regression of *y* on *x*

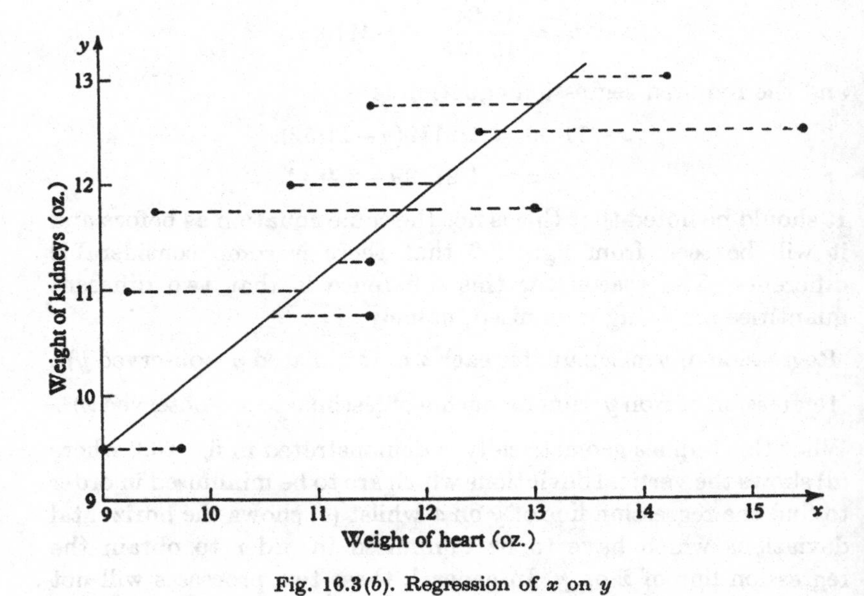

Fig. 16.3 (*b*). Regression of *x* on *y*

16.4 There is no need to carry out all the calculations for the regression lines in terms of the units of the original observations. Changes in scale and origin can be made and a conversion back to the original units carried out at the end. This will be illustrated with the short series of data in table 16.4.

Table 16.4. *Yields of roots and leaves of mangolds*

Plot no.	x	y	x'	y'	$(x')^2$	$(y')^2$	$(x'y')$
1	376·9	67·3	269	173	72,361	29,929	46,537
2	371·8	52·8	218	28	47,524	784	6,104
3	355·1	50·9	51	9	2,601	81	459
4	356·3	50·9	63	9	3,969	81	567
5	335·2	48·6	−148	−14	21,904	196	2,072
6	332·0	44·2	−180	−58	32,400	3,364	10,440
7	335·5	51·9	−145	19	21,025	361	−2,755
8	340·8	50·3	−92	3	8,464	9	−276
9	352·7	46·8	27	−32	729	1,024	−864
10	352·4	49·8	24	−2	576	4	−48
11	338·9	51·4	−111	14	12,321	196	−1,554
12	336·6	48·4	−134	−16	17,956	256	2,144

Example 16.2 In table 16.4, x is the yield in pounds of mangold roots whilst y is the yield in pounds of mangold leaves for the same plot. The data refer to twelve plots each of the same size. x' is obtained by taking the relation

$$x' = 10(x-350),$$

whilst y' is defined by the relation

$$y' = 10(y-50).$$

From the table the following totals are obtained:

$$\Sigma x' = -158, \qquad \Sigma y' = 133,$$

$$\Sigma(x')^2 = 241{,}830, \qquad \Sigma(y')^2 = 36{,}285,$$

$$\Sigma(x'y') = 62{,}826.$$

Hence the regression equation of y' on x' will be

$$y' - 11{\cdot}0833 = 0{\cdot}2694(x' + 13{\cdot}1667),$$

and for x' on y' will be

$$x' + 13{\cdot}1667 = 1{\cdot}8551(y' - 11{\cdot}0833).$$

These results can be used in this form throughout, or alternatively the equations can be converted back to the original units. In conversion back to the original units it should be noted that in this case both b and b' are unaltered. This is because the quantities involved are independent of origin and depend only on scale. As the scales of x and y have been changed in the same way, the alterations will be the same in both numerator and denominator and will thus cancel out, leaving b or b' unaltered. The modified equation for y on x will thus be

$$y-\tfrac{1}{10}(11{\cdot}0833)-50 = 0{\cdot}2694(x+\tfrac{1}{10}(13{\cdot}1667)-350),$$

or, simplifying

$$y-51{\cdot}1083 = 0{\cdot}2694(x-348{\cdot}6833),$$

and for x on y the regression line is

$$x-348{\cdot}6833 = 1{\cdot}8551(y-51{\cdot}1083).$$

16.5 In the great majority of practical problems only one of the two regression lines is required, but it is always important to decide which one it is. In the initial stages of an investigation both characteristics are measured in order to establish the form of the relationship between them. In subsequent work only one of the characteristics may be measured and the other estimated from it. In this case only one regression line is required and the other line need not be calculated.

The coefficients b and b' measure the slopes of the regression lines. If there is no relationship between the two characteristics the regression lines will be horizontal or vertical, as a knowledge of the value of one characteristic does not give any indication of the value of the other characteristic. Since the denominators of b and b' are variances, they are always positive; this implies that if b or b' is small, the numerator is also small. Hence the numerator, which is a symmetrical expression in x and y, is to some extent a measure of the relationship between the two characteristics. However, the numerator alone would not be a satisfactory measure, as it depends on the scale of the measurements. If all the values of x are multiplied by ten the numerator increases tenfold but the degree of relationship between x and y remains effectively the same. This makes it essential to introduce a factor that will take account of this undesirable property. The factor chosen is the standard deviation and the *coefficient of correlation* or association between two variables, x and y, is defined as

$$r = \frac{\Sigma xy - n\bar{x}.\bar{y}}{\sqrt{[\Sigma(x^2) - n\bar{x}^2]}\sqrt{[\Sigma(y^2) - n\bar{y}^2]}}. \tag{16.5}$$

This expression can be written in a number of ways, all algebraically equivalent, for example,

$$r = \frac{(1/n)\Sigma(x-\bar{x})(y-\bar{y})}{s_x s_y}, \tag{16.6}$$

since $$s_x^2 = (1/n)\Sigma(x^2) - \bar{x}^2.$$

It will be noticed that the denominator of r can only be positive but that the numerator can be either positive or negative. Whatever the values of x and y the values of r will always lie between $+1$ and -1. Positive values of r indicate positive relationships; that is, the higher the value of x the higher the value of y. Similarly, negative values of r indicate inverse relationships; that is, high values of one variable tend to be associated with low values of the other variable. Fig. 16.4 illustrates the situation with four scatter diagrams giving four different values of r. Each dot represents one pair of measurements and the changeover from negative correlation through zero to positive correlation is shown.

16.6 The calculation of the coefficient of correlation is straightforward. If the regression lines have previously been calculated then the coefficient may be obtained directly from the slopes since

$$r = \sqrt{(b \times b')}. \tag{16.7}$$

For the example on the weight of heart and weight of kidneys

$$b = 0{\cdot}3988, \qquad b' = 1{\cdot}2119$$

and $$r = \sqrt{(0{\cdot}4833)} = 0{\cdot}6592,$$

whilst in the example on the yields of roots and mangolds

$$b = 0{\cdot}2694, \qquad b' = 1{\cdot}8551$$

and $$r = \sqrt{(0{\cdot}4998)} = 0{\cdot}7070.$$

Both cases exhibit positive correlation, and in both cases the correlation is quite strong. This, of course, shows that given one of the two variates the other could be estimated from it with a fair degree of accuracy.

The calculation of the coefficient of correlation requires the same quantities as the calculation of the two regression lines. The method will be illustrated in example 16.3.

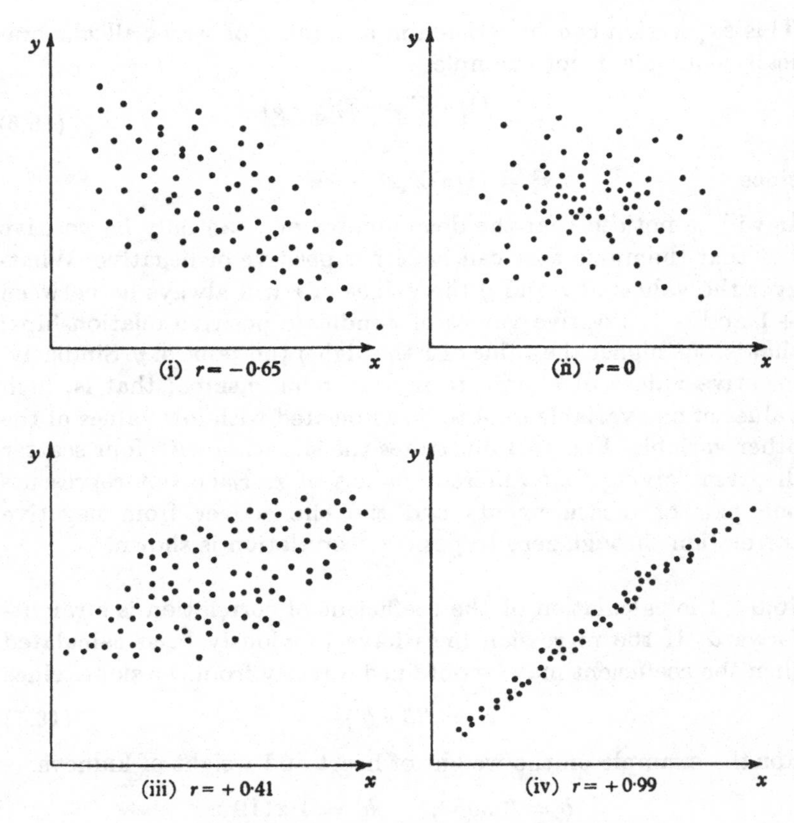

Fig. 16.4. Examples of scatter diagrams

Example 16.3 Twenty-seven candidates entered for a Civil Service examination. There were five compulsory subjects to be taken, the maximum mark in each subject being 300. In table 16.5 the marks, x, in arithmetic are given, together with the total marks, y, obtained in the other four subjects.

For the calculation an arbitrary origin was taken for x at 140, making $x' = x - 140$, and for y at 600, making $y' = y - 600$. The

Table 16.5. *Correlation of marks*

Candidate no.	x	x'	$(x')^2$	y	y'	$(y')^2$	$(x'y')$
1	230	90	8,100	907	307	94,249	27,630
2	218	78	6,084	748	148	21,904	11,544
3	187	47	2,209	677	77	5,929	3,619
4	186	46	2,116	658	58	3,364	2,668
5	182	42	1,764	698	98	9,604	4,116
6	167	27	729	643	43	1,849	1,161
7	164	24	576	824	224	50,176	5,376
8	162	22	484	725	125	15,625	2,750
9	158	18	324	683	83	6,889	1,494
10	154	14	196	746	146	21,316	2,044
11	151	11	121	645	45	2,025	495
12	150	10	100	628	28	784	280
13	141	1	1	580	−20	400	−20
14	139	−1	1	690	90	8,100	−90
15	138	−2	4	561	−39	1,521	78
16	135	−5	25	529	−71	5,041	355
17	130	−10	100	526	−74	5,476	740
18	126	−14	196	560	−40	1,600	560
19	124	−16	256	515	−85	7,225	1,360
20	113	−27	729	634	34	1,156	−918
21	101	−39	1,521	484	−116	13,456	4,524
22	90	−50	2,500	552	−48	2,304	2,400
23	78	−62	3,844	369	−231	53,361	14,322
24	71	−69	4,761	288	−312	97,344	21,528
25	61	−79	6,241	463	−137	18,769	10,823
26	48	−92	8,464	444	−156	24,336	14,352
27	37	−103	10,609	386	−214	45,796	22,042
Totals		−139	62,055		−37	519,599	155,233

various squares and cross products are then obtained and added up for the twenty-seven candidates giving

$$\Sigma x' = -139, \qquad \Sigma y' = -37,$$

$$\Sigma(x')^2 = 62{,}055, \qquad \Sigma(y')^2 = 519{,}599,$$

$$\Sigma(x'y') = 155{,}233.$$

Hence
$$\Sigma(x')^2 - n\bar{x}'^2 = 62{,}055 - 716 = 61{,}339.$$

$$\Sigma(y')^2 - n\bar{y}'^2 = 519{,}599 - 51 = 519{,}548,$$

$$\Sigma(x'y') - n\bar{x}'.\bar{y}' = 155{,}233 - 190 = 155{,}043.$$

Substitution in (16.5) gives

$$r = \frac{155{,}043}{\sqrt{(61{,}339)\,(519{,}548)}} = \frac{155{,}043}{178{,}518} = +0{\cdot}8685.$$

The denominator is always positive and the sign of r is the sign of the numerator. In this case there is a strong positive correlation between the two sets of marks, that is, candidates who get a high mark in arithmetic tend to get a high mark in the other subjects, and vice versa. Fig. 16.5 gives a scatter diagram of the marks and demonstrates the strong association between the two measurements.

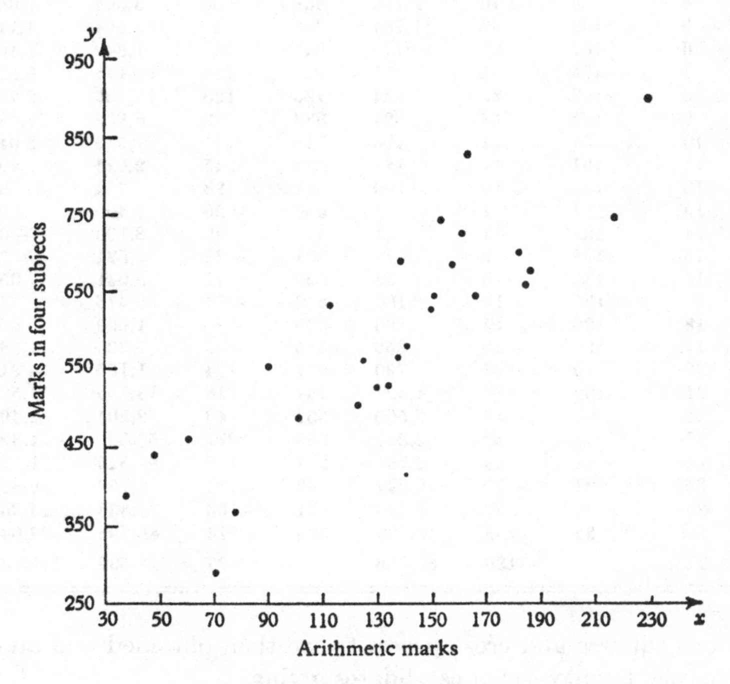

Fig. 16.5. Marks in examination

16.7 In many cases the data are not given, or are not available in the form of pairs of measurements, but in the form of a grouped two-way table such as the table for the heights and weights of schoolboys given at the end of chapter 3. In this case the technique is essentially the same, as each cell in the table is dealt with in turn and its contribution to the correlation calculated. The procedure is illustrated on the data in example 16.4, which contains rather more observations than would usually be met with in practice, but illustrates fully the principles of the calculation.

Table 16.6. *Length and breadth of laminae of runner-bean leaves*

Breadth of lamina in cm. (central values)

Length of lamina in cm. (central values)

y	x / y'	3·5	4·0	4·5	5·0	5·5	6·0	6·5	7·0	7·5	Totals
	x'	−4	−3	−2	−1	0	1	2	3	4	
8·5	5						1 *5*	1 *10*		1 *20*	3
8·0	4						1 *4*	2 *16*	5 *60*	2 *32*	10
7·5	3					3 *0*	4 *12*	16 *96*	12 *108*	2 *24*	37
7·0	2			1 *−4*		8 *0*	42 *84*	54 *216*	18 *108*	2 *16*	125
6·5	1		1 *−3*		9 *−9*	61 *0*	116 *116*	44 *88*	5 *15*		236
6·0	0			4 *0*	52 *0*	123 *0*	86 *0*	13 *0*	2 *0*		280
5·5	−1		1 *3*	20 *40*	76 *76*	75 *0*	18 *−18*	1 *−2*	1 *−3*		192
5·0	−2	1 *8*	4 *24*	28 *112*	39 *78*	15 *0*	2 *−4*				89
4·5	−3	2 *24*	4 *36*	8 *48*	5 *15*	1 *0*					20
4·0	−4	1 *16*	2 *24*	4 *32*	1 *4*						8
	Totals	4	12	65	182	286	270	131	43	7	1,000

Example 16.4 The data and workings are shown in table 16.6 and concern the length and breadth of laminae of runner-bean leaves. The data are grouped in intervals of width 0·5 cm. and central values are given. In each cell of the table the upper figure is the frequency of observation in that cell. The lower figure will be explained shortly. An arbitrary origin and scale are given to both x and y in order to make the computations more straightforward.

The new scales are

$$x' = 2(x-5{\cdot}5),$$

$$y' = 2(y-6{\cdot}0).$$

In any column of the table all the values of x' are equal and in any row of the table all the values of y' are equal. Hence the mean and variance of x' and y' can be found from the two margins which give the total number of observations in the respective columns or rows. These two marginal distributions both sum to 1,000, which is the number of observations in the whole table. Calculations give

$$\Sigma x' = 325, \qquad \Sigma(x')^2 = 1907,$$

$$\Sigma y' = 190, \qquad \Sigma(y')^2 = 2160.$$

For example,

$$\Sigma(y')^2 = 3\times 25+10\times 16+37\times 9+125\times 4+236\times 1+280\times 0$$
$$+192\times 1+89\times 4+20\times 9+8\times 16 = 2160.$$

Hence $\quad \Sigma(x')^2-n\bar{x}'^2 = 1907-105{\cdot}625 = 1801{\cdot}375,$

$$\Sigma(y')^2-n\bar{y}'^2 = 2160-36{\cdot}1 = 2123{\cdot}9.$$

The next step is to calculate the cross product term $\Sigma(x'y')$. In any one cell of the table all the observations will contribute the same amount to $\Sigma(x'y')$, namely the product of x' and y'. Hence the total contributions of the observations in that cell to $\Sigma(x'y')$ will be the value $x'\times y'$ for that cell, multiplied by the number of observations in the cell. This subsidiary calculation was made and the result recorded in italics as the lower number in each cell. For illustration consider the column headed 6·5.

First cell: $x' = 2$, $y' = 5$, number of observations 1, contribution $= 2\times 5\times 1 = 10$.

Second cell: $x' = 2, y' = 4$, number of observations 2, contribution $= 2\times 4\times 2 = 16$.

Third cell: $x' = 2$, $y' = 3$, number of observations 16, contribution $= 2\times 3\times 16 = 96$, and so on.

There is one row and one column of cells for which either x' or y' or both are equal to zero and these cells will make no contribution to $\Sigma(x'y')$. The remaining cells fall into four sections. In the upper right-hand and lower left-hand sections of the table the product $(x'y')$ is always positive, since x' and y' are either both positive or both negative. In the other two sections the product $(x'y')$ is always negative as one variable is positive and one is negative. It is convenient, therefore, to deal with each of the sections separately, and to add up the contributions to $\Sigma(x'y')$ from each of the four sections. These contributions are:

upper right-hand:	+1030,
lower right-hand:	− 27,
upper left-hand:	− 16,
lower left-hand:	+ 540.

Hence $\Sigma(x'y') = +1527.$

From (16.5)

$$r = \frac{\Sigma(x'y') - n\bar{x}'.\bar{y}'}{\sqrt{[\Sigma(x')^2 - n\bar{x}'^2]}\sqrt{[\Sigma(y')^2 - n\bar{y}'^2]}}$$

$$= \frac{1527 - 61{\cdot}75}{1956{\cdot}00} = 0{\cdot}7491.$$

As has been mentioned earlier no adjustments need be made to the coefficient to allow for the arbitrary scale and origin, since the coefficient is unaltered by any such changes. No adjustment is made for any error involved in assuming the frequencies to be concentrated at the mid-points of the intervals. Any errors in a cell due to this assumption will approximately cancel out with errors of opposite sign from other cells.

16.8 If there is no association or correlation between two variates sampling fluctuations may still cause a small sample to show a correlation that would not be found had a large sample of observations been available. Hence it is important to know how large a sample coefficient of correlation must be in order to be reasonably certain that correlation is present in the normal population from which the sample is drawn. As sampling fluctuations are of more magnitude in small than in large samples, a larger coefficient of correlation is necessary in a small sample to indicate the presence of correlation. The sampling distribution of r in samples drawn from a population with no correlation between the individuals has been obtained theoretically, and in table 16.7 the value of r necessary to establish significance at the 5 per cent level in a sample of size n is given. It should be noted that, to be significant, the observed value can either be less than $-r$ or greater than $+r$, and the larger the value of n the smaller that of r. In example 16.4 the coefficient of correlation was 0·75 from 1,000 pairs of measurements, and this is clearly a significant indication of association in the population from which this sample of observations has been drawn.

Table 16.7. *Significant values of r*

n	r	n	r
10	0·632	60	0·254
20	0·444	70	0·235
30	0·360	80	0·219
40	0·312	90	0·208
50	0·278	100	0·197

16.9 If it has been found that there is a significant degree of correlation between two measurements x and y, so that x can be used to estimate y by the equation

$$y-\bar{y} = b(x-\bar{x}),$$

then the next question is how efficient an estimator of y is obtained by this method. The answer to this lies in the variability of the observations about the fitted regression line. Suppose that in example 16.1, on the weights of heart and kidneys, the differences between the actual weights, y, and those estimated by the equation $y = 0{\cdot}3988x+6{\cdot}9338$ were found. These differences, or residuals as they are called, are a measure of the efficiency of estimation, since if they were all zero it would imply that the observed and estimated values of y coincided. To add up these residuals would be no use as a measure of the efficiency of the estimation, since some are positive, some are negative and their sum is approximately zero. If the variance of the residuals was used, the process of squaring eliminates the sign, and the larger the variance of the residuals the more the observations vary about the regression line. Let this variance of the residuals be denoted by s_e^2. Clearly what is needed is some method of calculating s_e^2 other than by the tedious process of evaluating, one by one, the differences between the estimated and actual values of y. This can be done by using the fact that the slope, b, of the regression line is dependent upon the values of x and y. If s_y^2 is the variance of the complete set of measurements of y, then as r increases from zero it is to be expected that s_e^2 will decrease, because the relationship between x and y removes some of the variability. The actual relationship, which will not be proved here, is

$$s_e^2 = s_y^2(1-r^2). \tag{16.8}$$

If r is zero the two variances coincide, which is to be expected as it implies that no knowledge is gained in the estimation of y by having the value of x available. If r is $+1$ or -1 then s_e^2 is zero because there is perfect association between the two variables and a knowledge of x fixes the value of y. The rapidity with which the overall variance, s_y^2, is reduced by correlating with a variable x is shown by the following figures:

$\pm r$	0·1	0·2	0·3	0·4	0·5	0·6	0·7	0·8	0·9
$1-r^2$	0·99	0·96	0·91	0·84	0·75	0·64	0·51	0·36	0·19

It will be seen that r has to be quite large before much reduction in the variance is obtained. For the weight of kidneys the value of s_y^2 is 1·2546 and r^2 is equal to 0·4833.

Hence

$$s_e^2 = 1{\cdot}2546(1-0{\cdot}4833) = 0{\cdot}6483,$$

or

$$s_e = 0{\cdot}8052.$$

This shows that a considerable increase in the accuracy of the estimation of kidney weight has been made by knowing the heart weight. For any x the estimated value of y will be $0{\cdot}3988x+6{\cdot}9338$ oz. and the standard deviation of this estimate is 0·8052 oz. If the observations were normally distributed about the estimate, some 95 per cent of the observations would be expected to fall on either side of the mean within a distance of 1·96 times the standard deviation. In this case the limits within which 95 per cent of the observations might be expected to fall are

$$(0{\cdot}3988x+6{\cdot}9338) \pm 1{\cdot}96 \times 0{\cdot}8052,$$

or

$$(0{\cdot}3988x+6{\cdot}9338) \pm 1{\cdot}5782.$$

Thus if x is equal to 11·50 oz. the appropriate limits for y would be 9·9418 and 13·0982 oz.

16.10 It must be emphasised that the coefficient of correlation and the regression lines are measures of the linear relationship between two variables, and that a low coefficient of correlation does not rule out the possibility that the variables are related in some other manner. The next example, example 16.5, gives a situation where the coefficient of correlation by itself does not adequately describe the data.

Example 16.5 The data in table 16.8 give the yield stress, x (that is the stress in tons/sq. in. beyond which Young's modulus does not hold) and the increase, in inches $\times 10^{-2}$, of the external diameter, y, of twenty steel tubes subjected to certain changes in internal pressure. The calculation of the coefficient of correlation follows the same pattern as before, after taking arbitrary scales and origins, and the necessary quantities are given in the table.

Table 16.8. *Data concerning steel tubes*

x	$x' = 10(x-20{\cdot}5)$	$(x')^2$	y	$y' = 10(y-7)$	$(y')^2$	$(x'y')$
20·0	−5	25	12·0	50	2,500	−250
20·0	−5	25	11·5	45	2,025	−225
19·0	−15	225	2·7	−43	1,849	+645
21·0	5	25	10·0	30	900	+150
22·2	17	289	7·8	8	64	+136
19·6	−9	81	9·2	22	484	−198
21·6	11	121	8·7	17	289	+187
19·1	−14	196	3·3	−37	1,369	+518
19·6	−9	81	7·3	3	9	−27
20·1	−4	16	10·7	37	1,369	−148
20·6	1	1	11·3	43	1,849	+43
21·4	9	81	11·2	42	1,764	+378
19·3	−12	144	4·0	−30	900	+360
19·1	−14	196	3·0	−40	1,600	+560
22·7	22	484	3·8	−32	1,024	−704
21·9	14	196	6·0	−10	100	−140
19·8	−7	49	10·8	38	1,444	−266
20·6	1	1	11·0	40	1,600	+40
21·1	6	36	10·8	38	1,444	+228
22·2	17	289	2·6	−44	1,936	−748
Totals	9	2,561		177	24,519	+539

$$\Sigma x' = 9, \qquad \Sigma y' = 177,$$
$$\Sigma(x')^2 = 2{,}561, \qquad \Sigma(y')^2 = 24{,}519,$$
$$\Sigma(x'y') = 539.$$

From these it follows, since n is equal to 20, that

$$\Sigma(x'y') - n\bar{x}'\bar{y}' = 459{\cdot}35,$$
$$\Sigma(x')^2 - n\bar{x}'^2 = 2556{\cdot}95,$$
$$\Sigma(y')^2 - n\bar{y}'^2 = 22{,}952{\cdot}55.$$

Hence, from equation (16.5)

$$r = \frac{459{\cdot}35}{\sqrt{(2556{\cdot}95)(22{,}952{\cdot}55)}} = 0{\cdot}0600.$$

The value of r is low, and for a sample of twenty is not indicative of significant correlation between the two variables. However, the mere calculation of r disguises an essential feature of the data which can be seen from an inspection of fig. 16.6. The figure shows

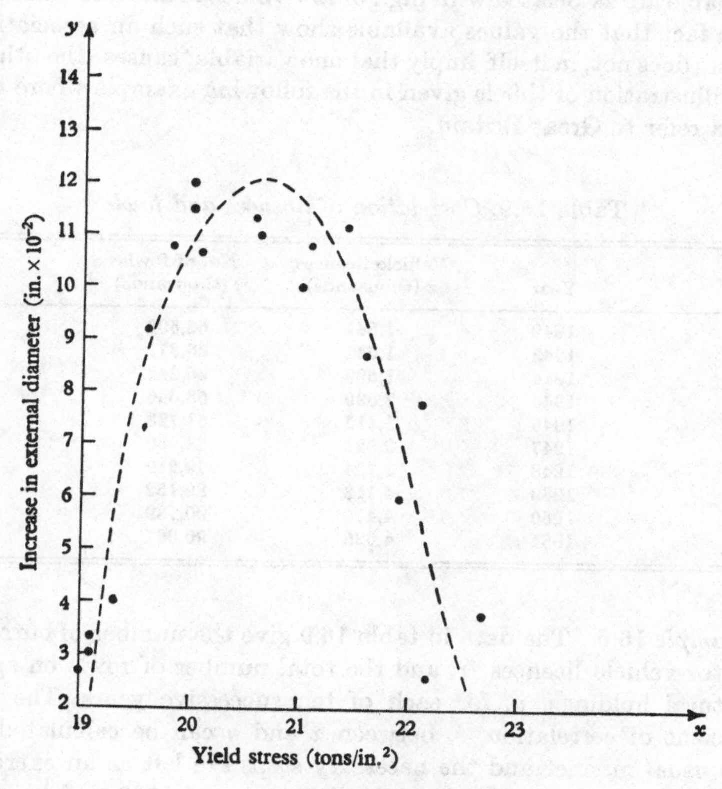

Fig. 16.6. Steel tube data

that there is a relationship between x and y, but that it is of a parabolic, not linear, form. The approximate nature of the relationship is indicated by the dotted line. All that the coefficient of correlation is doing is to indicate that the regression line of y on x is approximately horizontal whilst that of x on y is approximately vertical. As these two regression lines differ greatly the coefficient of correlation will be extremely low. This example shows that it is dangerous to calculate only the coefficient of correlation; some form of diagrammatic examination of the data should also be made in order to interpret the association between two measurements.

16.11 It must always be remembered that the coefficient of correlation merely measures the extent to which high values of one variable are associated with high or low values of another variable. The fact that the values available show that such an association exists does not, in itself, imply that one variable 'causes' the other. An illustration of this is given in the following example where the data refer to Great Britain.

Table 16.9. *Correlation of licences and fowls*

Year	Vehicle licences x (thousands)	No. of fowls y (thousands)
1942	1,847	53,502
1943	1,544	46,371
1944	1,599	50,242
1945	2,599	56,666
1946	3,113	61,723
1947	3,521	64,880
1948	3,734	79,219
1949	4,113	89,152
1950	4,414	90,789
1951	4,625	90,067

Example 16.6 The data in table 16.9 give the number of current motor-vehicle licences, x, and the total number of fowls on agricultural holdings, y, for each of ten successive years. The coefficient of correlation, r, between x and y can be calculated in the usual manner and the necessary steps are left as an exercise to the student. The final result gives $r = +0{\cdot}9566$ which is an extremely significant coefficient. It would be completely erroneous, however, to infer that one variable was the cause of the other, and that by banning motor-cars the population of fowls would immediately die off! The fallacy has arisen because both variables are in fact highly correlated with a third variable, time, and this is producing the spurious correlation that has been observed. Such relationships make the interpretation of correlation coefficients more difficult, and it is always essential to ask whether or not the two variables concerned do constitute cause and effect or not. The fundamental concepts involved in the study of regressions were put forward by Sir Francis Galton in the late nineteenth century when he was studying the relationships that existed in such physical phenomena as height and weight between parents and their children. It has, however, become a technique that is used in much wider fields, some of which have been mentioned in this chapter.

EXERCISES

16.1 The following table gives the marks, x, obtained by students at an examination in arithmetic at the end of one term together with the mark, y, obtained at the end of the following term.

Student	x	y	Student	x	y
1	53	41	7	47	45
2	74	65	8	72	59
3	48	44	9	48	20
4	71	38	10	65	57
5	66	41	11	80	64
6	60	62	12	40	27

(*a*) Draw a scatter diagram for these figures.

(*b*) Find the regression line for y on x and draw it on the scatter diagram.

(*c*) Determine the estimate of the coefficient of correlation between the two variables.

(*d*) What is your estimate of y for a student who obtains a mark of 60 for x, and what is the standard deviation of your estimate?

16.2 The percentage carotene content of wheat, x, and the percentage carotene content of the flour produced, y, for ten varieties of wheat were given by Goulden as follows:

Variety	x	y	Variety	x	y
1	1·18	2·39	6	1·25	1·76
2	2·13	3·11	7	1·65	2·10
3	1·41	2·15	8	1·24	2·12
4	1·42	1·96	9	1·48	2·28
5	1·50	2·02	10	1·35	1·86

(*a*) Draw a scatter diagram.

(*b*) Determine the regression line for y on x.

(*c*) Estimate the percentage carotene content of flour for x equal to 1·72 and give the standard error of the estimate.

16.3 Calculate the coefficient of correlation between the following series of male and female mortality rates per 1,000 of population:

Year	Male rate	Female rate	Year	Male rate	Female rate
1935	12·5	11·1	1940	16·1	12·9
1936	12·9	11·4	1941	15·7	11·8
1937	13·2	11·7	1942	14·4	10·7
1938	12·5	10·8	1943	15·3	11·3
1939	13·0	11·3	1944	15·3	10·8

16.4 An investigation was made to determine the manner in which the tensile strength of cement depends upon the curing time. The data are due to A. Hald, and after twenty-one experiments the following results were obtained for the five different times of curing used.

Curing time (days)	Tensile strength in kg./cm.^{2}
1	13·0, 13·3, 11·8
2	21·9, 24·5, 24·7
3	29·8, 28·0, 24·1, 24·2, 26·2
7	32·4, 30·4, 34·5, 33·1, 35·7
28	41·8, 42·6, 40·3, 35·7, 37·3

(*a*) Draw a scatter diagram of curing time, x, against tensile strength, y, and hence show that there does not appear to be a linear relationship between the two variables.

(*b*) Change to a new pair of variables $x' = 1/x$ and $y' = \log y$. Plot a scatter diagram of x' against y' and verify that there now appears to be a linear relationship between the two variables.

(*c*) Use the relationship in (*b*) to find the regression line of y' on x'.

(*d*) Make an estimate of the tensile strength that would result from a curing time of 15 days.

16.5 It might be expected that the price of crops varies inversely with the yield, that is, for a year with low yield the price is high and vice versa. The following figures give the price of oats and the yield per acre for ten years.

Year	Yield per acre (cwt.)	Average price *s.*	*d.*	Year	Yield per acre (cwt.)	Average price *s.*	*d.*
1940	17·0	14	0	1945	17·3	16	3
1941	16·4	14	7	1946	16·3	16	4
1942	17·2	15	1	1947	15·2	18	3
1943	16·7	15	9	1948	17·8	21	1
1944	16·2	16	4	1949	18·4	21	1

(*a*) Calculate the coefficient of correlation between yield and average price.

(*b*) Comment on the value of r obtained. Can you account for the value of r obtained, on grounds other than the supply and demand?

16.6 The following figures give the weight in grams, x, and the length of the right hind foot in millimetres, y, of a sample of twenty-eight adult male field mice.

x	y	x	y	x	y
15·4	22·6	17·4	22·2	18·8	21·5
16·0	22·2	17·6	22·4	18·8	22·5
16·0	22·6	17·8	22·1	18·8	23·2
16·5	22·4	17·9	22·8	19·2	23·0
16·9	23·5	18·2	22·5	19·4	22·4
16·9	23·3	18·4	22·4	20·1	23·5
16·9	21·8	18·6	23·0	20·4	23·3
17·0	22·2	18·7	22·9	20·4	23·4
17·2	21·9	18·7	23·9	22·3	23·0
17·4	21·9				

(*a*) Draw a scatter diagram for the data.

(*b*) Find the regression line of y on x and draw it in on the scatter diagram.

(*c*) Estimate the standard deviation, s_e, of the value of y found from the regression equation.

(*d*) Assuming that the observed values of y will be spread as in a normal distribution around the estimated y with a standard deviation of s_e, draw two lines parallel to the regression line such that 95 per cent of the observations might be expected to lie within the belt so formed.

16.7 The following figures give the average weekly output of coal in the United Kingdom and the number of civil servants for twelve successive quarters in the years 1950–3.

Output of coal (100,000 tons)	40	41	36	44	46	39	45	43	42	38	47	48
Civil servants (1000's)	689	685	679	676	675	680	686	688	684	678	673	668

(*a*) Calculate the coefficient of correlation between the two sets of figures.

(*b*) Would you agree that these figures demonstrate that a decrease in the number of civil servants results in an increase in coal production? Give reasons for your answer.

16.8 Four hundred and twenty-five children were given Binet intelligence tests and from the results the intelligence quotients of the children were calculated. After an interval of two years the children were re-tested and their revised intelligence quotients calculated.

(*a*) By drawing a graph with the means of the arrays of second test I.Q. against constant first test I.Q. see whether the regression is approximately linear.

(*b*) Calculate the regression line of I.Q. at second test on the I.Q. at the first test.

(*c*) With what accuracy could the I.Q. at the second test be predicted from a knowledge of I.Q. at the first test?

		I.Q. at first test (central values)						
		60	80	100	120	140	160	Totals
I.Q. at second test (central values)	160	—	—	—	2	5	6	13
	140	—	—	1	22	32	2	57
	120	—	1	26	79	15	—	121
	100	—	23	90	20	—	—	133
	80	2	58	24	1	—	—	85
	60	11	5	—	—	—	—	16
	Totals	13	87	141	124	52	8	425

16.9 A measurement, x, on the right claw and a measurement, y, of body length was made on each of 555 crabs.

(*a*) Calculate the regression of y on x. Estimate the body length of a crab for which x was equal to 6·21 mm.

(*b*) Calculate the coefficient of correlation between the two measurements.

		Measurement x (mm.) (central values)							
		4	5	6	7	8	9	10	Totals
Measurement y (mm.) (central values)	9	—	—	—	—	1	14	5	20
	8	—	—	3	11	86	63	2	165
	7	—	5	4	130	94	1	—	234
	6	2	6	52	55	—	—	—	115
	5	1	7	13	—	—	—	—	21
	Totals	3	18	72	196	181	78	7	555

16.10 It is frequently asserted that a seaside resort with a record of low rainfall has a high number of hours of sunshine. Investigate, with the aid of the following figures from Felixstowe, whether years with low rainfall have high sunshine by calculating the coefficient of correlation between the rainfall and the hours of sunshine.

Year	Rainfall (in.)	Sunshine (hr.)	Year	Rainfall (in.)	Sunshine (hr.)
1942	20·6	1690	1949	17·4	1907
1943	16·5	1951	1950	19·0	1629
1944	21·5	1755	1951	26·2	1714
1945	16·6	1623	1952	24·5	1727
1946	18·5	1687	1953	14·6	1521
1947	18·1	1851	1954	21·1	1491
1948	18·5	1669	1955	18·5	1636

16.11 The observations on p. 277 emerge from a surveillance inspection scheme in a controlled storage depot.

(*a*) Plot a scatter diagram.

(*b*) Calculate the least squares linear regression line for evaporation loss on storage time.

(*c*) Give 95 per cent confidence limits for the evaporation loss after a storage time of 15 days.

Aviation spirit (50 gallon drum)	Storage time (days)	Evaporation loss (pints)
No. 1	5	25
2	27	63
3	8	31
4	2	8
5	16	50
6	15	35
7	12	19
8	7	20
9	21	64
10	3	22

SOLUTIONS TO EXERCISES

CHAPTER 3

3.1 Frequencies, commencing with 0 group: 4, 12, 10, 22, 23, 21, 21, 7.

3.2 Frequencies: 8, 21, 18, 15, 7, 6, 2, 2, 0, 1.

3.3 (*a*) $\frac{1}{2}$, $1\frac{1}{2}$, 2, 0, $5\frac{1}{2}$, $9\frac{1}{2}$, $13\frac{1}{2}$, $5\frac{1}{2}$, 7, $9\frac{1}{2}$, $3\frac{1}{2}$, 2.
(*b*) 2, 2, 0. 4, 9, 15, 5, 6, 11, 3, 3.

3.4 (*a*) Frequencies, groups 149·5–161·5, etc.: 1, 3, 8, 13, 11, 6, 6, 2.
(*b*) Row averages: 182·5, 199·8, 194·2, 212·9, 208·3. It is probable that some, but not all, the litters were different in basic make-up.

3.5 Frequencies, groups 275–290, etc.: 2, 6, $10\frac{1}{2}$, 11, 7, $1\frac{1}{2}$, 1, 1.

3.6 Frequencies, groups 0·855–0·905, etc.: 6, 5, 6, 10, 13, 0, 2.

3.7 Frequencies, groups 15–20, etc.: 5, 3, 6, 20, $13\frac{1}{2}$, 21, 27, 12, $16\frac{1}{2}$, 1.

3.8 Frequencies, groups 10·25–11·45, etc.: 1,3, 2, 11, 14, 23, 23, 11, 10, 2.

3.9 Generally there is a low summer period, April to July, and a high autumn period, September to December. 1933 does not have the low period.

3.10 *Tensile strength (1000 lb. per sq. in., central values)*

		25·15	28·15	31·15	34·15	37·15	50·15	Total
Hardness (Rockwell's coefficient, central values)	54·25	5	5	3	1	—	—	14
	64·25	1	3	7	2	—	—	13
	74·25	1	4	3	10	1	—	19
	84·25	—	—	3	4	4	1	12
	94·25	—	—	—	—	1	1	2
Total		7	12	16	17	6	2	60

3.11 Frequencies, groups 1·5–, 2·0–, etc.: 1, 3, 8, 16, 14, 7, 2, 1.

CHAPTER 4

4.1 Possibilities are: bar chart, line diagram, block diagram, pie chart.

4.2 The main difference lies in the variation in the relative proportion of receipts coming from trams and trolleybuses.

4.4 (*a*) Line diagram. (*b*) 2,364.

4.8 Some arbitrary maximum has to be assumed. A figure of £100,000 would probably be reasonable.

4.9 (*a*) Block diagrams should be used either superimposed or else rotated through 90° to be on either side of the vertical axis. Note that, for practical purposes, the last groups can be assumed to have a maximum of 100. (*b*) The main difference is the rough equality of numbers at lower ages, but the increasing preponderance of females at the upper ages. (*c*) That the population in the age group 30–49 will increase markedly within 20 years.

4.12 Use a bar chart. Percentages are used and, for *A*, no total is given.

4.14 There is some evidence of an association between the two weights.

4.18 Bar charts are one method of presentation.

CHAPTER 6

6.1 Mean 2·157, mode 2.

6.2 O-level passes 1,036,457. A-level passes 176,525.

6.3

	Mean	Median
Heart	12·075	11·875
Kidneys	10·617	10·625
Span	69·6	70·15
Forearm	19·0	19·0

6.6 (*a*) 79·08. (*b*) 79·5. (*c*) 79·7 (obtained by splitting centre group inversely proportional to the groups immediately on either side of it).

6.7 (*a*) 172·2. (*b*) 170·0. (*c*) 171·9. This result is little effected by any reasonable assumptions that might be made regarding the end groups. Here it has been assumed that they have the same width as the other groups.

6.8 Mean, 0·9971; median, 0·9970; mode, 0·9970.

6.9 (*a*) Mode 2, median 3. (*b*) 3·36 to 3·39 by assuming all at 8 or alternatively spread from 8 to 12 with frequencies 4, 3, 2, 2, 1.

6.10 Mean, 39·69; median, 39·90; mode, 41. All ages are assumed to be age last birthday and mode has been estimated graphically.

6.11 Mean, 9·463; median, 9; mode, 8.

6.12 8·65 months.

6.13 Mean, 16·31, median, 16·30; mode, 16·3 (estimated graphically).

6.14 Mean, 0·579 to 0·580; median and mode 0.

6.15 Mean, assuming equal distribution within groups, ranges from £12,423 if upper limit is £60,000 for last group to £12,540 if upper limit is £75,000 for last group. Median, £8,713.

6.16 Mean, 2·52; median 2.

CHAPTER 7

(All calculations for standard deviations are based on a divisor of n, not $n-1$.)

7.1 (*a*) 29·47 years. (*b*) Males, 31·92; females, 35·32 years. (*c*) 43·22 lb. (*d*) 2 persons.

7.2 (*a*) 1·28 deaths. (*b*) 3·83 degrees. (*c*) 15·2 lb. (*d*) 16·5 lb. (*e*) 2·71 in. (*f*) 0·00114 in.

7.3 (*a*) 1·83 calls. (*b*) 15·95 words. (*c*) 1749 lb./sq.in. (1743 with Sheppard's correction). (*d*) 6·41 mm. (6·25 with Sheppard's correction). (*e*) 1·63 ounces. (*f*) 14·20 years (14·12 with Sheppard's correction).

7.4 (*a*) Mean deviation 0·905, standard deviation 1·902, ratio 1·21 (inches). (*b*) Mean deviation 3·24, standard deviation 3·73, ratio 1·15 (tensile strength 1000 lb./sq.in.). Mean deviation 9·94, standard deviation 11·07, ratio 1·15 (hardness). (*c*) Mean deviation 8·56, standard deviation 10·37, ratio 1·18 (years).

7.5 Mean deviation 4·69, standard deviation 5·81 mm.

7.6 Mean 36·0, standard deviation 8·40, mean deviation 6·81 years, ratio 1·23.

7.7 Mean 12·26, standard deviation 0·59 cm.

7.8 Mean 1·527, standard deviation 0·100 ounces.

7.9 Mean 21·30, standard deviation 8·53 years. **7.10** 1·80 seeds.

7.11 Mean 635·4, standard deviation 142 gallons.

7.12 (*a*) Mean 1062·9, standard deviation 101·5, \$1000. (*b*) Mean 1066·1, standard deviation 105·7, \$1000.

7.13 Mean 1·76, standard deviation 1·14 earners.

7.14 Standard deviation 586·7 gm., coefficient of variation 24·9 %.

7.15 Mean deviation 4·19, standard deviation 5·16, 0·0001 inches.

CHAPTER 8

8.1 (*a*) 5/18. (*b*) 1/6. (*c*) 1/12. **8.2** *A*'s chance 6/11, *B*'s chance 5/11.

8·3 (*a*) 4·7. (*b*) 3·7. The probabilities add up to 1 since the events are mutually exclusive and are the only two categories into which a digit may fall.

8.4 14. **8.5** 1/35.

8.6 12. **8.7** 1/5.

8.8 4/9. **8.9** (*a*) 1/4. (*b*) 1/52. (*c*) 1/26. (*d*) 2/13.

8.10 (*a*) 1/8. (*b*) 1/2. **8.11** 20.

8.12 18.

8.13 11/20.

8.14 (*a*) 1/3. (*b*) 8/27.

8.15 5/9.

8.16 Mean 27·1 and standard deviation 0·54.

8.17 n at least 17.

8.18 18·3 and 19·7 approximately.

8.19 0·3439.

CHAPTER 9

9.1 The probabilities of 0, 1, ..., 4 sixes are 625, 500, 120, 20, 1 divided by 1296. $x = 0$ is the most probable value.

9.2 The expected frequencies are: 0·9, 5·6, 14·1, 18·8, 14·1, 5·6, 0·9.

9.3 The probability of obtaining six or more bulls if each shot has a 0·4 chance of being a bull is 0·1662. This is not low enough for this hypothesis to be rejected in favour of there being a higher probability of getting a bull with each shot.

9.4 The probability of one or none breaking, with breakage rate of 25 per cent, is 0·0802. This is doubtful evidence that the breakage rate has declined.

9.5 The probability of 110 or less dying, with mortality rate of 0·247, is 0·089. This is not quite small enough for mortality rate to be rejected in favour of a lower one.

9.6 (*a*) 128/2187. (*b*) 448/2187. (*c*) 1/2187.

9.7 At least three.

9.8 Two boys has the highest probability. Mean 2·0, standard deviation 1·0.

9.9 (*a*) 19,683/262,144. (*b*) 59,049/262,144. (*c*) 183,412/262,144.

9.10 (*a*) 0·000729. (*b*) 0·117649.

9.11 The probability of 48 or more heads is 0·097.

9.12 The probability of 12 or more heads is 0·251.

9.13 The probability of 240 or more heads is approximately 0·00004. The difference is due to the bigger sample size.

9.14 63/256.

9.15 Expected frequencies: 0·7, 3·9, 10·5, 17·8, 21·3, 19·2, 13·5, 7·6, 3·5, 1·3, 0·6 over 9.

9.16 Expected frequencies: 77·5, 127·1, 83·3, 27·3, 4·5, 0·3. There is good agreement between observed and expected frequencies.

9.17 Expected frequencies, with a p of 0·45 estimated from the observed mean, are: 3·3, 13·6, 22·2, 18·2, 7·4, 1·2. There is reasonable agreement between observed and expected frequencies.

9.18 (*a*) 1/16. (*b*) ½.

CHAPTER 10

10.1 0·0361. **10.2** 0·0803. **10.3** 300.

10.4 (*a*) 0·223. (*b*) 0·191.

10.5 50, 150, 225, 224, 167, 100, 50, 21, 8, 3. **10.6** 0·112.

10.7 (*a*) 0·0168. (*b*) 0·156. (*c*) 0·859. **10.8** (*a*) 2:1:1. (*b*) 3:1:0.

10.9 0·9099. **10.10** 0·5. **10.11** $\frac{1}{3}$. **10.12** $S(0{\cdot}05+p)$.

10.14 (*a*) £0·002. (*b*) No. (Note 100p = £1.)

CHAPTER 11

11.1 No significant increase, $u = 0{\cdot}738$.

11.2 (*a*) 0·0455. (*b*) 0·379.

11.3 With a batch mean of 14·30, $u = 2{\cdot}670$ and the corresponding probability is 0·004. Hence such a batch stands a good chance of detection.

11.4 No significant increase, $u = 1{\cdot}155$.

11.5 No significant difference, $u = 1{\cdot}333$.

11.6 $u = 1{\cdot}827$. This is not quite significant at 5 per cent, two-tailed test.

11.7 $u = 3{\cdot}014$. Significant at 1 per cent.

11.8 No significant difference, $u = 0{\cdot}421$.

11.9 No significant difference, $u = 1{\cdot}067$.

11.10 $u = 3{\cdot}525$. Highly significant difference at 0·2 per cent.

11.11 No significant difference, $u = 0{\cdot}089$.

11.12 No significant difference, $u = 1{\cdot}399$.

11.13 $u = 2{\cdot}170$. Significant at 5 per cent, two-tailed test.

11.14 $u = 3{\cdot}605$. Significant at 0·2 per cent, two-tailed test.

11.15 No significant improvement, $u = 0{\cdot}995$.

CHAPTER 12

12.1 $\chi^2 = 21{\cdot}16$. Significant at 2·5 per cent.

12.2 No significant improvement, $\chi^2 = 9{\cdot}122$.

12.3 No significant increase, $\chi^2 = 9{\cdot}223$.

12.4 $u = 5{\cdot}397$. Highly significant increase.

12.5 $u = 2{\cdot}821$. Significant at 1 per cent.

12.6 $u = 3{\cdot}620$. Highly significant, so children were more variable.

12.7 0·7318 mm.

12.8 No significant difference from $\frac{1}{2}$, $u = 0{\cdot}413$.

12.9 $u = 1{\cdot}118$. Not significant at 10 per cent.

12.10 $u = 2{\cdot}297$. Significant at 5 per cent.

12.11 (*a*) 0·1579, standard deviation 0·032. (*b*) 1,330.

12.12 $u = 1{\cdot}943$. Significant difference at 5 per cent.

12.13 No significant effect, $u = 1{\cdot}001$.

12.14 Expected theoretical frequencies, with probability of a score of 5 equal to 1/9, are 331·5, 165·7, 31·1, 2·6, 0·1. These agree well with observations.

12.15 Approximately 2,300.

12.16 There is no significant difference so *A* could be retained.

12.17 Estimated value of *p*, 0·24. Expected frequencies are: 4·8, 15·2, 21·6, 18·2, 10·1, 1·0. Agreement with the observations is good.

12.18 $t = 1{\cdot}72$. Not significant at 5 per cent.

12.19 $t = 0{\cdot}74$. Not significant at 5 per cent.

CHAPTER 15

15.1 Experimenting with periods ranging from 6 to 11 suggests that 10 is the most suitable. The trend values (starting at 1886) are then: 5·3, 5·5, 5·9, 6·1, 5·9, 5·3, 4·8, 4·5, 4·4, 4·4, 4·4, 4·3, 4·0, 3·8, 3·7, 3·7, 3·7, 4·0, 4·5, 4·9.

15.2

Year	Month ...	1	2	3	4	5	6
1943		101·7	102·1	102·4	102·6	102·8	103·0
1944		103·5	103·6	103·6	103·7	103·8	103·9
1945		104·9	105·1	105·2	105·3	105·5	105·7

Year	Month ...	7	8	9	10	11	12
1942		99·0	99·5	100·4	100·4	100·9	101·3
1943		103·2	103·2	103·4	103·4	103·4	103·4
1944		104·1	104·2	104·3	104·5	104·6	104·8

15.3 Trend values (commencing at 1948, 7th month) are: 50·9, 51·0, 50·9, 50·7, 50·8, 50·8, 51·1, 51·3, 51·6, 52·0, 52·0, 52·0, 51·9, 51·8, 52·0, 52·0, 51·8, 51·9, 51·8, 51·4, 51·0, 50·8, 50·5, 50·1, 49·9, 49·6, 49·3, 49·1, 48·8, 48·7, 48·7, 48·7, 48·8, 48·9, 49·4.

The seasonal oscillations (commencing at first month) are: −9·4, −8·7, −7·4, −3·0, +1·9, +8·9, +11·0, +10·6, +8·1, +1·6, −5·5, −9·5.

The random oscillations (commencing at 1948, 7th month) are: −1·3, −1·6, −1·1, −0·9, +1·1, +2·2, +0·9, +0·6, −2·0, +1·9, −0·9, −1·4, +0·8, +0·8, +2·3, +0·8, −1·3, +1·3, −1·7, −0·2, +2·6, −1·3, +0·6, +2·3, +0·4, +0·7, −1·2, 0·0, +0·1, −3·4, +0·7, −0·3, −0·5, −0·5, +0·2, −0·8.

Random oscillations: mean = 0·0, standard deviation = 1·4.

15.4 Trend values (commencing at third quarter, 1952) are: 98·2, 97·5, 98·8, 98·9, 99·1, 99·1, 98·7, 98·4, 98·8, 99·5, 99·9, 101·4.

The seasonal oscillations (commencing at first quarter) are: +12·0, −13·5, +21·1, −19·8.

The random oscillations (commencing at third quarter, 1952) are: −1·0, +2·7, −3·2, +2·1, +0·6, −0·4, −0·3, +0·1, +0·3, −2·4, +3·4, −2·3.

Using a graph the best estimate for the second quarter of 1956 is 86·5. The actual number is 80·9.

15.5 The seasonal oscillations (commencing at January) are: +2·3, −0·7, −1·4, +0·9.

15.6 October 1953, 6083; January 1954, 7503; July 1956, 5601.

15.7 Experimenting with a moving average of different periods, it seems that a six-point average is most satisfactory. From this it appears that there is a slight rise over the early period but no marked tendency to increase or decrease over the remainder.

15.8 (*a*) Trend values (commencing at third quarter 1952) are: 2399, 2449, 2510, 2576, 2642, 2713, 2794, 2879, 2962, 3050, 3149, 3252, 3350, 3446, 3540, 3626. (*b*) The seasonal oscillations (commencing at first quarter) are: −178, +61, +95, +10. (*c*) Mean 0, standard deviation 19·9.

15.9 (*a*) 9·5 per cent. (*b*) 9·2 per cent. Generally (*b*) would be preferable since a constant percentage rate of increase produces a linear trend whereas (*a*) gives a curved trend.

15.10 (*c*) −1·4, −2·1, 6·4, −2·8.

CHAPTER 16

16.1 (*b*) $y = 0{\cdot}809x - 1{\cdot}89$. (*c*) 0·710. (*d*) Estimate 46·65, standard deviation 10·86.

16.2 (*b*) $y = 1{\cdot}025x + 0{\cdot}677$. (*c*) Estimate 2·44, standard deviation 0·27.

16.3 0·465. **16.4** (*c*) $y' = -0{\cdot}4975x' + 1{\cdot}6017$. (*d*) 37·0.

16.5 (*a*) 0·432. (*b*) Inflation and price restrictions due to the war may have affected true relationship.

16.6 (*b*) $y = 0{\cdot}1566x + 19{\cdot}8176$. (*c*) 0·536. **16.7** (*a*) −0·432.

16.8 (*b*) $y = 0{\cdot}896x + 11{\cdot}959$. (*c*) Approximately ±18·6.

16.9 (*a*) $y = 0{\cdot}7154x + 1{\cdot}7721$. Estimate 6·21. (*b*) 0·866.

16.10 −0·292. **16.11** (*b*) $y = 2{\cdot}14x + 8{\cdot}9$. (*c*) 20 to 50.

BIBLIOGRAPHY

The short list below gives a few books that could be studied to follow up the material presented in this introductory text. The first eight books are general statistical books whilst the next three deal with more specialised fields of application of statistical methods. Finally an inexpensive, but extremely useful, short set of statistical tables is listed.

(1) *The Nature of Statistics* by W. Allen Wallis and Harry V. Roberts (Collier Macmillan 1962).

(2) *Elementary Statistics* by P. G. Hoel (Wiley, 4th edition, 1976).

(3) *Introductory Statistics* by T. H. Wonnacott and R. J. Wonnacott (Wiley, 3rd edition, 1978).

(4) *Basic Statistics: A Modern Approach* by Morris Hanbury (London; Harcourt Brace Jovanovich, 1974).

(5) *Statistical Analysis* by E. C. Bryant (McGraw Hill, 2nd Edition, 1966).

(6) *Statistical Methods in Management* by T. Cass (Cassell, 2nd edition, 1973).

(7) *The Anatomy of Decisions* by P. G. Moore and H. Thomas (Penguin, 1976).

(8) *Introduction to Statistical Inference* by E. S. Keeping (Van Nostrand, Reinhold, 1963).

(9) *Statistics for Economists* by Sir Roy Allen (Hutchinson University Library, 1966).

(10) *Applied General Statistics* by F. E. Croxton, D. J. Cowden and S. Klein (Pitman, 3rd edition, 1968).

(11) *Statistical Theory with Engineering Applications* by A. Hald (Wiley, 1965).

(12) *Cambridge Elementary Statistical Tables* by D. V. Lindley and J. C. F. Miller (Cambridge University Press, 1965).

INDEX